TECHNICAL REPORT

T0146342

Unconventional Fossil-Based Fuels

Economic and Environmental Trade-Offs

Michael Toman, Aimee E. Curtright, David S. Ortiz,
Joel Darmstadter, Brian Shannon

Sponsored by the National Commission on Energy Policy

Environment, Energy, and Economic Development

A RAND INFRASTRUCTURE, SAFETY, AND ENVIRONMENT PROGRAM

This research was sponsored by the National Commission on Energy Policy and was conducted under the auspices of the Environment, Energy, and Economic Development Program (EEED) within RAND Infrastructure, Safety, and Environment (ISE).

Library of Congress Cataloging-in-Publication Data

Unconventional fossil-based fuels : economic and environmental trade-offs / Michael Toman ... [et al.].
    p. cm.
  Includes bibliographical references.
  ISBN 978-0-8330-4564-5 (pbk. : alk. paper)
  1. Petroleum engineering. 2. Heavy oil. 3. Oil sands. 4. Coal liquefaction. I. Toman, Michael A.
II. RAND Corporation.

TN871.U49 2008
333.79'68—dc22

                                                                      2008036873

The RAND Corporation is a nonprofit research organization providing objective analysis and effective solutions that address the challenges facing the public and private sectors around the world. RAND's publications do not necessarily reflect the opinions of its research clients and sponsors.

RAND® is a registered trademark.

Published 2008 by the RAND Corporation
1776 Main Street, P.O. Box 2138, Santa Monica, CA 90407-2138
1200 South Hayes Street, Arlington, VA 22202-5050
4570 Fifth Avenue, Suite 600, Pittsburgh, PA 15213-2665
RAND URL: http://www.rand.org/
To order RAND documents or to obtain additional information, contact
Distribution Services: Telephone: (310) 451-7002;
Fax: (310) 451-6915; Email: order@rand.org

Preface

Rising concerns about energy costs and security, as well as about greenhouse-gas (GHG) emissions from use of petroleum-based motor fuels, have stimulated a number of public and private efforts worldwide to develop and commercially implement alternatives to petroleum-based fuels. Commonly considered fuel options for the medium term (roughly 10–20 years) include both biomass-based fuels (e.g., ethanol, biodiesel) and unconventional fossil-based liquid fuels derived from such sources as heavy oils, oil sands, coal liquefaction, and oil shale.

This report assesses potential future production levels, production costs, GHG emissions, and environmental implications of unconventional fossil-based motor fuels derived from oil sands and coal. The study was sponsored by the National Commission on Energy Policy as part of a larger body of sponsored research to investigate the portfolio of options needed to address cost, energy-security, and GHG concerns about motor fuels. The report is intended to be of use to policy analysts and decisionmakers concerned with each of these aspects of motor fuels, as well as to the general public that will confront the economic and environmental implications of different policy choices in this arena.

This study builds on earlier RAND Corporation studies on natural resources and energy development in the United States. Most relevant are the following:

- *Producing Liquid Fuels from Coal: Prospects and Policy Issues* (Bartis, Camm, and Ortiz, forthcoming)
- *Oil Shale Development in the United States: Prospects and Policy Issues* (Bartis, LaTourrette, et al., 2005)
- *Understanding Cost Growth and Performance Shortfalls in Pioneer Process Plants* (Merrow, Phillips, and Myers, 1981).

The RAND Environment, Energy, and Economic Development Program

This research was conducted under the auspices of the Environment, Energy, and Economic Development Program (EEED) within RAND Infrastructure, Safety, and Environment (ISE). The mission of ISE is to improve the development, operation, use, and protection of society's essential physical assets and natural resources and to enhance the related social assets of safety and security of individuals in transit and in their workplaces and communities. The EEED research portfolio addresses environmental quality and regulation, energy resources and systems, water resources and systems, climate, natural hazards and disasters, and economic development—both domestically and internationally. EEED research is conducted for government, foundations, and the private sector.

Questions or comments about this report should be sent to the project leader, David Ortiz (David_Ortiz@rand.org). Information about EEED is available online (http://www.rand.org/ise/environ). Inquiries about EEED projects should be sent to the following address:

Debra Knopman, Director, ISE
Environment, Energy, and Economic Development Program, ISE
RAND Corporation
1200 South Hayes Street
Arlington, VA 22202-5050
703-413-1100, x5667
Debra_Knopman@rand.org

Contents

Figures

Tables

Summary

Background

Both the price of petroleum motor fuels and concerns regarding emissions of carbon dioxide (CO_2) are driving attention to possible substitutes. In 2008, the world price of oil reached record highs after being adjusted for inflation, continuing a pattern of price increases over several years. Petroleum products derived from conventional crude oil constitute more than 50 percent of end-use energy deliveries in the United States and more than 95 percent of all energy used in the U.S. transportation sector. Emissions from the consumption of petroleum account for 44 percent of the nation's CO_2 emissions, with approximately 33 percent of national CO_2 emissions resulting from transportation-fuel use (EIA, 2007a). Commonly considered alternative transportation-fuel options for the near and medium terms (roughly 10–20 years) include both biomass-based fuels (e.g., ethanol, biodiesel) and unconventional fossil-based liquid fuels derived from such sources as heavy oils, oil sands, oil shale, and coal liquefaction.

In this report, RAND researchers assess the potential future production levels, production costs, greenhouse gases (GHGs), and other environmental implications of synthetic crude oil (SCO) produced from oil sands and transportation fuels produced via coal liquefaction (often referred to as *coal-to-liquids* [CTL]). Production of liquid fuels from a combination of coal and biomass is also considered. Although oil shale is also an important potential unconventional fossil resource, we do not address it in this report because fundamental uncertainty remains about the technology that could ultimately be used for large-scale extraction, as well as about its cost and environmental implications. The omission from this report of renewable fuel options and other propulsion technologies should not be interpreted as a conclusion that the fossil-based options are superior to others.[1]

Policy Context

Concerns about high oil prices reflect not only the hardships endured by many energy users but also the large transfer of national wealth to foreign oil producers (particularly members of the Organization of the Petroleum Exporting Countries [OPEC]) that are widely perceived to elevate prices above competitive market levels by restricting output. Such artificially elevated oil prices provide a rationale for policy intervention to support the production of alternative fuels, because increased competition from alternative sources limits petroleum exporters' abil-

---

[1] For further information about renewable options, see Toman, Griffin, and Lempert (2008); Bartis, LaTourrette, et al. (2005) provided a detailed analysis of oil shale.

ity to influence the market. In addition, sudden oil-price spikes are widely seen to have adverse effects on national employment and output levels. Alternative fuels may reduce the instability of oil prices by lowering the potential size and likelihood of sudden reductions in crude-oil supply. However, the magnitude of the effect on short-term market instability is likely to be small so long as the alternative fuels make up only a relatively small increment in world fuel production. Accordingly, we focus in this report on the longer-term price effects.

There also are increasing concerns about the adverse impacts of climate change from rising global emissions of GHGs. CO_2 is the most important GHG in terms of total volume and impact on the climate, and most CO_2 emissions result from fossil-fuel use. According to the Intergovernmental Panel on Climate Change (IPCC, 2007), increased accumulation of CO_2 and other heat-trapping gases in the earth's atmosphere is likely (albeit with varying degrees of quantitative uncertainty) to change the climate in a variety of ways, with a variety of adverse effects. While the U.S. share of global emissions (currently about one-quarter) will decline as energy use in the developing world continues to grow rapidly over the next few decades, the Energy Information Administration (EIA) projects that U.S. emissions will rise by about one-third between 2007 and 2030, with emissions from transportation maintaining their roughly one-third share of this larger total (EIA, 2007a).

Within this broader context of concern for energy cost and security and CO_2 emissions, the possibility for increasing use of liquid fuels derived from oil sands and coal raises several specific questions. One set of questions concerns the potential production volumes of these alternative fuels (since this will affect the size of benefits from increased competition with crude oil) and the potential production costs of these fuels (since this will influence their competitiveness in the market and thus their ability to provide such competition for conventional crude oil–based products). Another set of questions concerns the potential life-cycle emissions of CO_2 from these substitutes relative to conventional fuels and the relative costs of mitigating increased emissions from transportation fuels. These sets of economic and environmental questions are linked by the fact that the future unit costs of alternative fuels in the market will depend on advances in their technologies and the costs of addressing their CO_2 emissions; the competitiveness of the alternative fuels will depend on the potential future price of crude oil and the cost of addressing CO_2 emissions from conventional fuels.

Technical Approach

For both SCO and CTL, we provide a bottom-up assessment of potential future production, potential costs, and potential environmental and other barriers to capacity expansion. The environmental barriers addressed include CO_2 emissions and more local and regional concerns related to water and land. Our primary focus is on the longer term, although we also discuss the issues that arise in ramping up capacity over the intervening period. Production of SCO is already occurring on a significant scale in Canada, using several technologies. CTL, on the other hand, is produced only to a limited extent on a commercial scale in South Africa, so its analysis is based on studies of how modern technology might perform if deployed in the United States. In addition, we discuss the use of capture and geological storage of CO_2 emissions resulting from the production of the two alternative energy sources. Carbon capture and storage (CCS) consists of separating out CO_2 emissions then transporting them to sites where they can be injected deep underground for long-term storage. The added cost of CCS is the cost, including return on investment, for capture, transportation, and storage.

We then investigate how three key drivers influence the future cost-competitiveness of fuels from SCO and CTL relative to fuels from conventional crude oil:

- the future price of crude oil
- changes in the unit production costs of the unconventional fossil-based fuels induced by further technical advances and experience in their production
- the implications of potential constraints on CO_2 emissions for the unit production costs of both conventional and unconventional fossil-based fuels.

The future course of each unit-cost driver is uncertain, so we compare the fuels under a number of plausible scenarios to represent the key uncertainties. EIA's 2007 *Annual Energy Outlook* (AEO) (EIA, 2007a, Table 12) has a reference-case price of light sweet crude oil in 2025 of about $56/barrel (bbl) (in 2005 dollars), while the high-oil-price case reflects a 2025 price of about $94/bbl. (The low-price case is about $35/bbl.) The costs of production of the technologies also are uncertain. For oil sands, new extraction technologies are being brought forward whose future costs are uncertain. For coal liquefaction, there is not yet experience with modern plant designs implemented on a larger scale. Finally, we consider ranges of CCS costs and potential costs of fuel supply from future regulations to limit CO_2 emissions.

It is difficult to estimate future production costs for unconventional fuels. There is often a bias toward underestimating costs and overestimating performance of new fuel-production facilities and their operations. Since facilities that upgrade and refine bitumen from oil sands or produce CTL require significant levels of investment, the average cost of producing a unit of product over the facility's lifetime is sensitive to a number of assumptions regarding the time to construct the facility, the mixture of capital and debt used to finance the construction, the costs of the feedstocks, and the successful start-up and long-term capacity factor of the facility. All of these parameters are uncertain and difficult or impossible to accurately predict early in the planning process. We attempt to account for some of this uncertainty by providing ranges of cost estimates for recovering bitumen from oil sands and for coal liquefaction. There are opportunities for significant improvements in production costs as experience is gained. A first-of-a-kind plant may be subject to significant cost overruns and poor performance, but subsequent plants may resolve these issues and perform significantly better. Taking these considerations into account, for the year of interest (2025), we derived low and high cost estimates for the production of SCO and CTL.

To account for how costs associated with limiting CO_2 emissions may affect SCO and CTL competitiveness with respect to conventional petroleum or fuels, we incorporate a complete life-cycle-emission analysis of each fuel. Life-cycle emissions are those associated directly and indirectly with primary production of feedstock, processing, transporting, and, ultimately, the use of the end product, including gasoline, diesel fuel, or close unconventional substitutes for these.

We address the impacts of potential limits on CO_2 cost-competitiveness in two ways. In scenarios in which we assume that CCS does not occur, the cost of CO_2 emissions is a measure of the increased cost of supplying and using each fuel due to future regulatory constraints on CO_2 emissions from production and final use of the fuel. The life-cycle emissions per unit of fuel times the cost of CO_2 emissions released to the atmosphere is added to our estimated production cost of fuel, to arrive at a cost that includes the effects of CO_2-emission constraints. In no-CCS scenarios, we can highlight the sensitivity of cost-competitiveness to production costs, and we establish a basis for evaluating the potential competitiveness of CCS investment. When CCS is an option, the added cost associated with potential future CO_2 constraints is the cost per unit of CO_2 captured and stored times the quantity of stored CO_2 plus the cost of CO_2 emissions (described earlier) applied to noncaptured emissions. Fuel producers will apply CCS when its unit cost is less than the cost of CO_2 emissions released to the atmosphere.

Key Findings

Basic production costs for SCO are likely to be cost-competitive with conventional petroleum fuels. Production of SCO already is a relatively mature technology, though new processes are being developed to make use of deeper formations. Taking into account both uncertainties that may lead to higher costs than estimated and cost improvements due to learning, and leaving aside for the moment the potential cost of CO_2 emissions, we find that SCO is cost-competitive with conventional petroleum unless future oil prices are well below EIA's 2007 reference-case scenario for 2025.

While basic production costs for CTL also appear to be competitive with conventional petroleum fuels across a range of crude-oil prices, **CTL competitiveness is more sensitive to technology costs and to oil prices.** In the absence of a CO_2-emission cost, CTL fuels appear to be competitive with conventional petroleum fuels if oil prices are above the EIA 2007 reference-case price in 2025. However, if CTL turns out to be more costly than anticipated or oil prices in the longer term are lower than this reference price, CTL may not be cost-competitive even without a CO_2-emission cost.

Higher oil prices or significant energy-security premiums increase the economic desirability of SCO and CTL. If longer-term oil prices are high or future energy-security policy attaches a high premium to the market price of crude oil to account for energy-security costs, then investment in both SCO and CTL production will be correspondingly more favorable. In particular, the range of CO_2-emission costs over which CTL without CCS is still economically attractive relative to conventional diesel will increase, and the economics of CTL with CCS can look attractive relative to conventional petroleum even if CCS turns out to be relatively costly. On the other hand, if oil prices end up being relatively low over the longer term, then CTL is less competitive than petroleum, even with a low CO_2-emission cost.

Even with future policy constraints on CO_2 emissions and their associated costs, SCO seems likely to be cost-competitive with conventional petroleum; the main potential constraint on SCO production is its local and regional impacts. SCO is only about 15–20 percent more CO_2-intensive on a life-cycle basis than conventional crude, even without CCS, and has essentially the same CO_2 intensity with CCS. Therefore, its potential cost advantages relative to future oil prices are maintained over a wide range of potential CO_2 emission–control costs. For oil sands, the prominent limiting factors appear to be the high water usage that would accompany a major scaling up of SCO production, attendant concerns about water quality, other environmental impacts and socioeconomic constraints, and (to a lesser extent) the availability of natural gas for bitumen extraction and upgrading.

The cost-competitiveness of CTL is more dependent than that of SCO on the costs of CO_2 emissions and CCS. If CCS can be deployed on a large scale and at a relatively low cost, then CTL with CCS appears to be economically competitive over a wide range of conventional-fuel prices and CO_2-emission costs. The picture would change only if long-term oil prices were significantly lower than the 2007 EIA reference-case value. However, if CCS and CTL costs end up being relatively high, then CTL is cost-competitive with conventional fuels at EIA's high price for 2025, but not at the reference-case price. Other constraints on CTL production could include environmental concerns associated with increased coal mining and the availability of water for CTL plant processes.

Unconventional fossil fuels do not, in themselves, offer a path to greatly reduced CO_2 emissions, though there are additional possibilities for limiting emissions. Fuels derived from oil sands and CTL emit fossil-based CO_2 during combustion, just as conventional petroleum products do. Thus, even when employing CCS to capture and store CO_2 emitted during fuel production, life-cycle emissions of CO_2 for these alternative fuels are comparable to those of conventional fuels. Large-scale production of these unconventional fuels does not reduce emissions of CO_2. Reliance on liquefaction of a mixture of coal and biomass along with CCS does have the potential to achieve greatly reduced life-cycle emissions, but potential production of such fuels would be limited by the availability and cost of the biomass feedstock and the potential availability and cost of CCS.

Relationships among the uncertainties surrounding oil prices, energy security, CCS costs, and CO_2-control stringency have important policy and investment implications for CTL. Our analysis indicates that investment in CCS for CTL can be a very robust undertaking if CCS can be realized at an adequately large scale, if CTL and CCS costs are in the lower part of the range of costs that we have considered, and if future oil prices do not fall below reference-case levels. If CTL and CCS costs are higher, however, CCS's value to the CTL supplier as a hedge against the cost of future CO_2 controls is positive only with higher long-term (not just near-term) oil prices.

From a societal perspective, it is desirable to reduce the need for rapid and costly CO_2-emission reductions through implementing a less abrupt approach to CO_2 limits. It is also desirable to take actions that increase the availability of cost-effective alternatives to conventional petroleum. If nearer-term concerns about energy security lead to emphasis on rapid CTL investments while CO_2-control requirements are delayed or kept minimal, then energy-security and climate-protection objectives are brought into conflict.

Neither CTL investors nor policymakers have many options for reducing long-term oil-price uncertainty. As noted, moreover, there is a risk to the economic value of CTL investment just from the possibility of relatively low long-term prices. On the other hand, policymakers do have options for reducing the uncertainties surrounding CTL and CCS costs. There is a large benefit from government financing for continued research and development (R&D) for CCS *and* initial CCS investments at a commercial operating scale to further assess the technical and economic characteristics of CCS. This analysis parallels the argument in Bartis, Camm, and Ortiz (forthcoming) for active but limited public-sector support for informative initial-scale investment in modern CTL facilities. Conversely, it may be very beneficial socially to delay a significant ramp-up in CTL production until the uncertainties surrounding CCS technology and CTL-production costs can be reduced. These observations reflect the importance of the argument of the National Commission on Energy Policy (NCEP) (2004) for a broad portfolio of technology-development initiatives and a variety of policy instruments to promote energy diversity and decarbonization of fuel sources.

Acknowledgments

The authors gratefully acknowledge advice and assistance from a number of current and former RAND colleagues, including James T. Bartis, Raj Raman, and Nathaniel Shestak, and from several members of the National Commission on Energy Policy staff, including Sasha Mackler, Nate Gorence, and Tracy Terry. The report was considerably strengthened thanks to careful and detailed comments offered by individuals at the National Energy Technology Laboratory, Natural Resources Defense Council, the Pembina Institute, and Rentech. None of these individuals bears responsibility for any remaining errors in the report.

Abbreviations

AEO	Annual Energy Outlook
API	American Petroleum Institute
bbl	barrel
bbl/d	barrels per day
bcf/d	billion cubic feet per day
Btu	British thermal unit
CAPRI	catalytic method developed in part by the Petroleum Recovery Institute
CBTL	coal and biomass to liquid
CCS	carbon capture and storage
CERI	Canadian Energy Research Institute
CH_4	methane
CO	carbon monoxide
CO_2	carbon dioxide
CO_2e	carbon-dioxide equivalent
CSS	cyclic steam stimulation
CTL	coal-to-liquids
dilbit	diluted bitumen
DVE	diesel value equivalent
EEED	Environment, Energy, and Economic Development Program
EIA	Energy Information Administration
EOR	enhanced oil recovery
FEED	front-end engineering design
FT	Fischer-Tropsch

gal.	gallon
GDP	gross domestic product
GHG	greenhouse gas
GREET	Greenhouse Gases, Regulated Emissions, and Energy Use in Transportation
GTL	gas to liquid
GW	gigawatt
GWP	global-warming potential
H_2	hydrogen
ICO_2N	Integrated CO_2 Network
IEO	International Energy Outlook
IGCC	integrated gasification combined cycle
IPCC	Intergovernmental Panel on Climate Change
IRR	internal rate of return
ISE	RAND Infrastructure, Safety, and Environment
kWh	kilowatt-hour
LPG	liquefied petroleum gas
Mcf	thousands of cubic feet
mmBtu	millions of British thermal units
MTG	methanol to gasoline
Mton	megaton
MW	megawatt
N_2O	nitrous oxide
NEB	National Energy Board
OPEC	Organization of the Petroleum Exporting Countries
PC	pulverized coal
PPI	producer price index
PRI	Petroleum Recovery Institute
psia	absolute pounds per square inch
R&D	research and development
SAGD	steam-assisted gravity drainage

SCO	synthetic crude oil
SFC	Synthetic Fuels Corporation
SOR	steam-to-oil ratio
synbit	bitumen blended with synthetic crude oil
THAI	toe-to-heel air injection
VAPEX	vaporized extraction
WTI	West Texas Intermediate

CHAPTER ONE

Introduction

Background

Petroleum products derived from conventional crude oil constitute 55 percent of end-use energy deliveries in the United States and more than 95 percent of energy used in the U.S. transportation sector. Although less CO_2-intensive per British thermal unit (Btu) than coal, oil-derived liquids account for 44 percent of the nation's carbon-dioxide (CO_2) emissions.[1] There also are concerns about the long-term cost of petroleum-based energy, the economic and other implications of large wealth transfers to oil exporters, and price instability in petroleum markets.[2]

These rising concerns about both energy security and greenhouse-gas (GHG) emissions from use of petroleum-based motor fuels have stimulated a number of public and private efforts worldwide to develop and commercially implement alternatives to conventional petroleum-based fuels. A major focus in the near term has been improving fuel economy, both in the aggregate and through increased penetration of hybrid electric vehicles. The most commonly considered alternative fuel options for the medium term (roughly 10–20 years) are biomass-based fuels (e.g., ethanol, biodiesel) and unconventional fossil-based liquid fuels derived from heavy oils, oil sands, coal liquefaction, and oil shale, as well as advanced plug-in electric hybrids. In the longer term, hydrogen (H_2) may also emerge as a solution, although this fuel currently faces many more fundamental technical hurdles than the other options mentioned here.

In this report, we assess the potential future production levels, production costs, GHG emissions, and other environmental implications of two fossil-based alternative fuels. These are fuels derived from bitumen extracted from oil sands and fuels produced by conversion of coal to liquid fuels. The first is often called *synthetic crude oil* (SCO), while the second is often referred to as *coal-to-liquids* (CTL). Production of liquid fuels from a combination of coal and biomass is also briefly considered. Although oil shale is also an important potential unconventional resource, we do not address it in this report because fundamental uncertainty remains about the technology that could ultimately be used for large-scale extraction, its costs, and environmental implications. The omission from this report of renewable fuel options should not be interpreted as a conclusion that the fossil-based options are superior.[3]

---

[1] See EIA (2007c, Tables 1.3, 2.1b–2.1f, and 10.3). A summary chart can be found at EIA (2007d).

[2] For historical as well as contemporary context on energy-security concerns and misconceptions, see Parry and Darmstadter (2003).

[3] Each option has costs and benefits, and each must be weighed and compared to others on that basis. For further information about renewable options, see Toman, Griffin, and Lempert (2008); Bartis, LaTourrette, et al. (2005) provided a detailed analysis of oil shale.

In the absence of measures to capture and permanently store CO_2, SCO from oil sands and CTL, will have higher CO_2 emissions/unit of fuel than will conventional fuels. The feasibility and costs of limiting or offsetting these higher emissions is one critical consideration in evaluating these fuels. Ultimately, it is likely that a portfolio of options will be needed to address cost, energy-security, and GHG concerns from motor fuels. This report considers one part of such a portfolio-wide approach.

Technical Approach

For both SCO and CTL, we first provide a bottom-up assessment of potential future production, potential production costs, and potential technical and environmental barriers to capacity expansion. Environmental barriers may include both CO_2 emissions and local or regional impacts of resource development. Our primary focus is on the longer term, although we also discuss the issues that arise in ramping up capacity in the intervening period. We also provide a similar review of the technological and economic aspects of carbon capture and storage (CCS) in geological formations.

We then investigate how three key drivers influence the future cost-competitiveness of fuels from oil sands and CTL relative to fuels from conventional crude oil:

- the future price of crude oil
- changes in the unit costs of the unconventional fossil-based fuels induced by further technical advances and experience in their production
- the implications of potential constraints on CO_2 emissions for the unit costs of both conventional and unconventional fossil-based fuels.

The future course of each unit-cost driver is uncertain, so we compare the fuels under a number of plausible scenarios to represent the key uncertainties. The Energy Information Administration's 2007 *Annual Energy Outlook* (EIA 2007a, Table 12) has a reference-case price of light sweet crude oil in 2025 of about $56/barrel (bbl) (in 2005 dollars), while the high-oil-price case reflects a 2025 price of about $94/bbl. (The low-price case is about $35/bbl.) The costs of production of the technologies are also uncertain. For oil sands, new extraction technologies are being brought forward whose future costs are uncertain. For coal liquefaction, there is not yet experience with more-modern plant designs. Finally, we consider ranges of future CCS costs and potential costs imposed by future regulation to limit CO_2 emissions.

It is difficult to estimate future production costs for unconventional fuels. There is often a bias toward underestimating costs and overestimating performance of new fuel-production facilities and their operations. Since facilities that upgrade and refine bitumen from oil sands or produce CTL require significant levels of investment, the average cost of producing a unit of product over the lifetime of the facility is sensitive to a number of assumptions regarding the time to construct the facility, the mixture of capital and debt used to finance the construction, the costs of the feedstocks, and the successful start-up and long-term capacity factor of the facility. All of these parameters are uncertain and difficult or impossible to predict accurately early in the planning process. We attempt to account for some of this uncertainty by providing a range of cost estimates for recovering bitumen from oil sands or coal liquefaction. There are opportunities for significant improvements in production costs as experience is gained. A first-of-a-kind plant may be subject to significant cost overruns and poor performance, but subsequent plants may resolve these issues and perform significantly better. For the year of interest (2025), we derived low and high cost estimates for the production of SCO and CTL.

To account for how costs associated with limiting emissions of CO_2 may affect the competitiveness of SCO and CTL with respect to conventional petroleum or fuels, we incorporate a complete life-cycle-emission analysis of each fuel. Life-cycle emissions are those associated directly and indirectly with primary production of feedstock, processing, transporting, and, ultimately, the use of the end product, including gasoline, diesel fuel, or close unconventional substitutes for these. We consider a range of assumed values for a cost of CO_2 emissions associated with some form of regulation to limit CO_2. For our purposes, the form of regulation does not need to be specified.

We address the impacts of potential CO_2 limits on cost-competitiveness in two ways. In scenarios in which we assume that using CCS is not possible, the cost of CO_2 emissions applies to all emissions associated with production and final use of the unconventional fuels and conventional petroleum, based on their life-cycle CO_2 intensity. The cost of CO_2 emissions then is added to each fuel's basic supply cost. We consider such no-CCS scenarios in order to be able to highlight the sensitivity of cost-competitiveness to fuel production–cost uncertainties, as well as to provide a basis for comparison to identify the potential impact of CCS on cost-competitiveness.

When the option of CCS is included in the cost-competitiveness analysis, the added cost/ unit of energy is the cost for applying CCS to a portion of emissions, plus the cost of CO_2 emissions applied to noncaptured emissions (including again the emissions from final use of the fuel). Fuel producers will implement CCS when its cost/unit of CO_2 stored is less than the cost/unit of CO_2 emissions released.

Organization of This Report

Chapter Two provides some historical context on synfuel development and key background on the energy-security and GHG concerns motivating interest in the alternative fuels, particularly unconventional fossil-based ones. Chapters Three through Five review the particulars of CCS, SCO from oil sands, and fuels from CTL. Chapter Six examines the cost-competitiveness of SCO and CTL fuels relative to conventional petroleum under different assumptions about technology, crude-oil prices, and CO_2 storage and emission costs. Chapter Six also addresses the implications of incorporating monetized values of energy-security costs. Chapter Seven summarizes the study and provides some broader conclusions.

CHAPTER TWO

History and Context of Unconventional Fossil-Resource Development

Past U.S. Efforts to Promote Synfuels

The Synthetic Fuels Corporation (SFC) was a U.S. government–sponsored program to develop a capacity to produce synthetic fuels in the early to mid-1980s. Critics of new efforts to promote unconventional fossil fuels often use the poor results of SFC as an argument for keeping the government out of the role of alternative energy–resource development. Our brief review of the SFC experience in this chapter is intended to highlight cautionary lessons and to indicate how current circumstances differ from those of SFC.

Dramatic oil-price increases due to world oil-market upheavals in the 1970s gave special impetus to the creation of SFC as a public but quasi-independent institution under the Energy Security Act of 1980 (P.L. 96-294). Much of the momentum driving the formation and financing of SFC rested on the prospect of synfuel costs being within a range likely to be approached and even surpassed by world oil prices within a near-term planning horizon. Against this backdrop, the goal was to stimulate production of shale oil and coal-derived fuels through a variety of financial incentives.[1]

By 1987, production was expected to be no less than 500,000 barrels per day (bbl/d). By the early 1990s, it was expected that a synfuel-production capacity of several million bbl/d would be likely, albeit with prospective federal financial support in the billions of dollars. Technological obstacles and the need for a commercial learning curve to reduce production costs were scarcely considered. Moreover, such optimism was voiced not only by SFC's federal backers. A nongovernmental panel of experts, eyeing the production target of the equivalent of 1.75 million bbl/d for shale oil and liquefied coal by 1990, characterized the required technologies as "ready for deployment [needing only] financial incentives to proceed to production" (U.S. House of Representatives, 1980).

These expectations proved to be short-lived, however. The world oil-price collapse in the mid-1980s (with a two-thirds decline from 1981 to 1986) eliminated the possibility of achieving anything close to SFC's objectives. The government closed SFC in 1986, even though its enabling law had called for termination between 1992 and 1997.

Among the lessons that have been drawn from the SFC experience is that government, notwithstanding its undeniably important role in supporting research on innovative energy systems, ought to be wary of targeting specific resources or technologies. However, a broader

---

[1] Congress passed the first Synthetic Liquid Fuels Act in 1944 (58 Stat. 189), providing to the U.S. Department of the Interior $455 million authority in loan guarantees for synfuel development. Some level of coal liquefaction or gasification research and development (R&D) has been ongoing continuously since the 1940s. For a detailed discussion of the formation of SFC and its goals, see Schurr et al. (1979).

lesson is relevant to the current debate about developing unconventional fuels. While it is hard to determine whether technological hurdles alone—apart from the world oil-price collapse—would have been enough to sink the synfuel efforts of the 1980s, the fact that oil prices did not trend inexorably higher (as expected) offers an important caution in current assessments of unconventional-fuel potential. Moreover, environmental factors may be significant deterrents to the expansion of unconventional fossil fuels.

Energy Information Administration Production Projections

Production projections by the Energy Information Administration (EIA) provide a standard set of commonly accepted reference numbers that can be used to compare production of conventional and unconventional fuels under different scenarios. They thus provide a useful point of departure for other assessments of future production. EIA's 2007 *International Energy Outlook* (IEO) projected the contribution of conventional and unconventional fossil resources to the liquid-fuel supply.[2] The contribution of oil sands is projected to rise from approximately 1 million bbl/d in 2007 to 3.6 million or 4.4 million bbl/d in 2030 (depending on whether one considers the reference case or high-oil-price scenario). Projected CTL production in the United States is shown in Table 2.1. CTL output is more limited in the near term, and it is higher in the high-oil-price case than in the reference case, given its improved cost-competitiveness in that case. Shale oil (not included in Table 2.1) materializes only in the high–world oil-price scenario.

Potential Sources of Oil-Sand and CTL-Capacity Investment

There are a number of major players in current oil sand–production efforts, including Suncor Energy, the original company to make SCO from oil sands in 1967 (see Suncor, undated), and Syncrude Canada, a consortium of major oil companies, including ConocoPhillips and Exxon Mobil (Imperial Oil in Canada) (see Canadian Oil Sands Trust, undated; Syncrude, undated). Each company has a production level on the order of 350,000 bbl/d. Canadian Natural Resources and Petro-Canada are also major oil sand–production companies. The Athabasca Oil Sands Project is "one of the largest construction projects on the planet" and is a joint venture between Royal Dutch Shell (which acquired Shell Canada Ltd), Chevron Canada Resources (a wholly owned subsidiary of Chevron), and Western Oil Sands (see Albian Sands,

Table 2.1
EIA (2007) CTL Output Projections

Year	Reference Case (thousands of bbl/day)	High-Price Case (thousands of bbl/day)
2015	100	140
2020	100	600
2025	300	1,200
2030	400	1,600

SOURCE: EIA (2007b, Tables G.3 and G.6).

_____

[2]   See EIA (2007b, Tables G.3 and G.6). These were the most recent projections available when the research was undertaken. The 2008 IEO indicated a higher oil-price trajectory.

undated). Additionally, many companies involved in joint ventures are looking to make significant independent expansions as well, including ConocoPhillips (Surmont in situ project), Exxon Mobil/Imperial Oil (Cold Lake in situ and Kearl mining projects), and Royal Dutch Shell. Smaller but still consequential projects include those of Devon Energy, Nexen, and OPTI Canada. Most of these companies are producing, or planning to produce, more than 100,000 bbl/d. Other companies poised to pursue oil-sand production on a significant scale are North American Oil Sands (recently acquired by Statoil ASA, the Norwegian state oil firm) and Total E&P Canada Ltd.[3]

Today, Sasol Limited in South Africa operates the world's only commercial CTL production. It currently produces the energy equivalent of about 160,000 bbl/d of fuels and chemicals (Steynberg, 2006; Sasol, 2006). According to a 2004 worldwide survey, at least 13 new facilities based on coal gasification began operations between 1993 and 2004 (NETL, undated) and are still operating today. That survey also listed an additional 25 facilities that would begin operations during 2005 and 2006. These coal-gasification facilities produce synthesis gas (syngas), a mixture of carbon monoxide (CO) and H_2. Most of the facilities produce syngas for use in the manufacture of chemicals, and six facilities are dedicated to producing electric power using a combination of gas and steam turbines that is often referred to as an *integrated gasification combined cycle* (IGCC). Table 5.1 in Chapter Five contains more information about these projects. Coal-gasification facilities, whether for chemicals or power, involve much the same operations as would be required at the front end of a modern Fischer-Tropsch (FT) CTL plant.

Policy Drivers for Unconventional Fossil-Based Fuels: Greenhouse-Gas Emissions and Energy Security

Concerns About Greenhouse Gases

To properly compare the overall emissions of CO_2 from different fuels, it is necessary to evaluate the emissions of CO_2/comparable unit of energy across the entire life cycle of the fuel, or the *full fuel cycle*. This means calculating the emissions associated directly and indirectly with primary production of feedstock, as well as the processing, transportation, and, ultimately, the use of the end product, whether it be gasoline, diesel fuel, or close unconventional substitutes for these. In the absence of measures to control CO_2 emissions, the potential future emissions of CO_2 associated with producing SCO and CTL fuels are higher/unit of product than are those of conventional petroleum.

An evaluation of the implications of greater use of unconventional fuels needs to account for this greater CO_2 intensity/unit of energy supplied and how it might affect the relative competitiveness of the unconventional fuels. The growing attention being paid to limiting GHG emissions in the United States and globally could lead to various forms of regulatory constraints on CO_2 emissions into the atmosphere. Those constraints, in turn, would lead to added costs for different fuels based on their GHG intensity. At this stage, it is not possible to predict the form or stringency of future limits on CO_2 emissions in the United States or the extent to which these limits would affect motor fuels. For this reason, we represent the impact of future

---

[3] Useful information can be found in Alberta Employment, Immigration and Industry (2007). An up-to-date inventory of existing and proposed Canadian oil-sand activities can be found in Strategy West (2008).

CO_2 controls parametrically in our unit-cost comparisons by allowing for different values of the cost of CO_2 emissions and different costs for CCS.

Concerns About Energy Security

Since the oil-price shocks of the 1970s, there has been persistent concern about the adverse economic consequences of both high and unstable oil prices.[4] We review these concerns at a general level and then put them in the context of transportation fuels derived from conventional and unconventional fossil resources.

The concern about high oil prices reflects not just the resulting burdens on individual energy users. It also reflects the national implications of large transfers of national wealth to foreign oil producers—in particular, members of the Organization of the Petroleum Exporting Countries (OPEC) that many observers see as holding prices above competitive market levels by restricting output. Artificially elevated oil prices provide a rationale for policy intervention, including policies to stimulate production of cost-competitive alternative fuels to mitigate exporters' use of market power.[5] Even a small oil-price reduction accruing to consumers over a large volume of oil consumption and imports can add up to a significant economic benefit.[6]

While any fuel substitution has the potential to lower international oil prices by reducing demand for conventional petroleum, the magnitude of the benefit will depend on the cost-effectiveness of the alternative fuels and their production potential. Substitution of considerably more expensive fuels through various possible measures, such as subsidies or fuel-use mandates, erodes the economic benefits gained from a lower world oil price and reduced wealth transfer. Moreover, the degree of oil cost savings will depend on oil producers' responses (Bartis, Camm, and Ortiz, forthcoming). For example, if they reduce their output to buffer the decline in oil prices, the import cost savings would be weakened as well.

Oil-price spikes also are a concern because of their adverse impacts on national employment and output. The specific mechanisms behind these adverse, economy-wide impacts remain subjects of research and debate, but they are generally seen to result in lower employment when reduced energy use lowers the marginal product of labor in the economy. While alternative transportation fuels might reduce the instability of oil prices by lowering the potential size and likelihood of price shocks, this benefit is likely to be quite small unless unconventional fuels make up a large share of total demand. With a small market share, the prices of the substitutes will be highly correlated with prices for conventional petroleum products. In this report, we focus on the potential benefit of alternative fuels in terms of lower long-term oil prices and smaller international wealth transfers.

---

[4] For recent discussions of these issues, see Huntington (2005) and Leiby (2007).

[5] The reduction in wealth transfer affects all foreign oil suppliers, not just OPEC. Domestically, the effect of lower oil prices is to transfer economic surplus from oil producers to oil consumers.

[6] Policies to reduce fuel demand through improved energy efficiency can also yield this benefit.

CHAPTER THREE

Carbon Capture and Storage for Unconventional Fuels

This chapter presents an overview of technology and costs for CCS as it relates to extracting bitumen from oil sands and producing liquid fuels from coal. The *capture of CO_2* refers to methods of isolating a concentrated stream of CO_2 and pressurizing it in preparation for transportation by pipeline to permanent storage. *Storage of CO_2* refers to permanent, belowground storage of CO_2. Significant research, development, and demonstration are under way in the United States and throughout the world to identify sites that would support large-scale, permanent, geologic storage of the CO_2. Several large-scale tests are under way.

The systems and processes for capturing CO_2 in oil sand–extraction and –upgrading facilities and in CTL facilities are commercially proven, and systems for transporting and injecting CO_2 are in widespread use today. We can use this experience to derive cost estimates for CCS for SCO and CTL. CO_2 capture differs in some important ways between bitumen extraction and upgrading and CTL production, as discussed in this chapter. Additional detail regarding the quantity of CO_2 captured during operations and the actual costs of capture are presented in the following chapters for each of the technologies individually. Costs of transportation and storage are considered in this chapter to the extent that there are similarities for the two technologies. The only cost that is assumed to be identical on a per-unit basis for both technologies is that of storage. The presentation here is brief; the interested reader is encouraged to refer to the cited documents for more detail.

Carbon-Dioxide Capture

Centralized facilities offer the best opportunities for CO_2 capture (IPCC, 2005). For oil sands, the various point sources associated with facilities that extract bitumen and upgrade the recovered bitumen (e.g., boilers, heaters, on-site power generators) are the points at which it is most convenient to capture CO_2. For CTL, removing CO_2 from key process streams is part of normal operations, so capture principally involves preparing the CO_2 for pipeline transport (Bartis, Camm, and Ortiz, forthcoming). For this reason, the marginal cost of CCS per unit CO_2 captured is higher for SCO than for CTL.

In oil-sand operations, trucks, excavators, and other fleet equipment do not offer a practical opportunity for CO_2 capture, although these are more significant sources of emissions for mining operations than for in situ ones. Additional sources, such as mine-face and tailing-pond emissions (for mining) and fugitive gases and flaring emissions (for in situ operations), are also relatively difficult to capture. These are the nonpoint sources of GHGs that we assume will not be subject to CCS.

It is less costly to capture CO_2 emissions from large, stationary combustion sources and bitumen-upgrading facilities. For example, current methods for in situ production of oil sands require the on-site production of the steam that is injected to promote the flow of the bitumen to a producing well. In our analysis, we assume that the cost of capture from all such combustion sources in oil-sand operations will be comparable to capture costs for a new pulverized-coal (PC) power plant, as estimated by the Intergovernmental Panel on Climate Change (IPCC).[1] Similarly, we apply IPCC costs for CO_2 capture in H_2-production facilities to the capture of such emissions from oil sand–upgrading facilities. In both cases, rather than using the entire (very broad) range of values presented by IPCC, we instead use the representative cost values therein, with a small variation around these values to account for uncertainty and site variability.[2] These costs include the cost of pressurizing the captured CO_2. We then assume a 25-percent decrease in these costs by 2025, consistent with IPCC's assessment of the prospects for technological improvements in capture technologies during this time frame. More details are presented in Chapter Four.

Isolation of a pure stream of CO_2 is an integral process in CTL facilities employing the FT and methanol-to-gasoline (MTG) processes (see Chapter Five). In these facilities, the coal is gasified under high temperature and pressure to produce syngas. One by-product of this process is a waste stream of CO_2, which is removed from the syngas as part of preparing to produce the liquid fuels. In some FT plants, a second stream of CO_2 is removed during the synthesis process. The technology for isolating CO_2 from these process streams is well established and commercially proven (Bartis, Camm, and Ortiz, forthcoming). Ten to 20 percent of the plant-site CO_2 emissions are produced in the section of the CTL plant that generates electric power. Technical detail regarding the capture of CO_2 in CTL facilities is provided in Chapter Five.

Once isolated, the final step in CO_2 capture is the pressurization of the CO_2 in preparation for transport. This step would be common to both oil-sand and CTL facilities employing systems for CO_2 capture. The stream of CO_2 needs to be pressurized to at least 1,200 absolute pounds per square inch (psia) (IPCC, 2005) but typically in the range of 2,000 to 2,200 psia (SSEB, 2005, Appendix D) to allow for pressure losses during transport and to drive the geologic disposal process. Electricity produced on site is used to operate the compression equipment. In our analysis, the cost of CO_2 capture is estimated by including the capital and operating costs of the compression equipment in the financial analysis.

Carbon-Dioxide Transport

The captured and pressurized CO_2 must be transported from the capture site to the storage site. This is typically performed via pipeline. There is considerable commercial experience in the pipeline transport of CO_2 in North America dating to the 1970s and 1980s. Kinder

---

[1]   IPCC (2005). It is also possible, in theory, to convert the fossil fuel used in these various point sources into H_2 and CO_2; the H_2 would be burned and the CO_2 captured. This is a plausible option because H_2 is also required for other plant operations, principally the upgrading of the bitumen to SCO. However, in this analysis, we do not assume that this will be standard practice.

[2]   We use the representative values in IPCC ± $5/ton CO_2. This not only narrows the very broad range presented by IPCC, but it also centers the distribution on this representative value, which is not the central value of the ranges presented by IPCC.

Morgan operates a 1,300-mile pipeline network in the United States for use in enhanced oil recovery (EOR) (Kinder Morgan, 2006). Pipelines in the western United States have a capacity to transport more than 50 million tons of CO_2 per year from both natural reservoirs and built sources to EOR operations (IPCC, 2005). Most relevant to the topic of CCS for both oil sands and CTL is the 205-mile CO_2 pipeline connecting a coal-gasification plant in North Dakota with the Weyburn, Saskatchewan, CO_2 EOR and storage test site. This pipeline delivers approximately 2 million tons per year of CO_2 at a pressure of 2,200 psia (IPCC, 2005).

Pipeline and infrastructure costs for CO_2 transport are proportional to the distance the CO_2 must be transported and to the size of the CO_2-generating facility. The only difference in our analysis between oil sands and CTL with respect to CO_2-transportation costs is the difference in the quantity of CO_2 produced and captured on a daily basis.[3] For both technologies, we assume transport costs as estimated by IPCC for a 250-km pipeline but at different total production volumes (IPCC, 2005, Figure 8.1). As will be discussed later, a moderately sized CTL facility will capture approximately 50 percent more CO_2 than a moderately sized oil sand–extraction and –upgrading operation. Because CO_2 transport is mature technology, we do not assume any cost improvements by the year 2025.

Carbon-Dioxide Storage

EOR and geologic storage are the two options currently being considered for disposing of captured CO_2 emissions from unconventional-fuel production. EOR is the only commercial option currently available for the disposition of appreciable CO_2 emissions. In general, EOR activities have not been intended to provide permanent storage. Given favorable geology, however, certain sites have the potential to permanently store CO_2 (NETL, 2008). For oil-sand and CTL plants built in the near term, EOR would provide an opportunity to dispose of captured CO_2. However, we are interested in long-term, large-scale options for CO_2 beyond the potential scale of storage through EOR.

Substantial efforts are under way worldwide to develop dedicated geologic storage for CO_2. Additional options for disposing of plant-site CO_2 emissions may become available in the longer term. These include the use of CO_2 in facilities dedicated to biomass production (e.g., algae farms). The costs of identifying, commissioning, decommissioning, and long-term monitoring of storage sites are uncertain and the focus of significant U.S. and international study (NETL, 2007b; IPCC, 2005).

Oil-sand and CTL facilities would utilize common technologies for CO_2 storage without obvious technical distinctions or cost differences. In this report, we use the same values for storage costs, which include monitoring and verification, for both technologies—namely, the entire range of values given by IPCC for geologic storage (see IPCC, 2005, Table 8.2). Because of the noted cost uncertainties, we do not make any assumptions about the cost of geologic storage decreasing by the year 2025.

---

[3] The geographic distribution of oil-sand operations may, in fact, make capturing and transporting a significant fraction of CO_2 more complicated and, therefore, more expensive than this analysis implies. For example, we have assumed collocation of in situ and upgrading operations herein. However, increasing CCS costs for oil sands would only strengthen the conclusions in later chapters with respect to incentives to use CCS in the oil-sand industry.

Enhanced Oil Recovery

In EOR operations, CO_2 is used to increase production in certain oil fields in a method known as *CO_2 flooding*. In this process, CO_2 is injected into a field at pressure, where it mixes with the remaining oil in the field, changing its flow properties. The oil may then be pumped from a producing well. Typically, much of the injected CO_2 is recovered from the EOR site and reinjected or used elsewhere, though a certain amount of CO_2 remains in the reservoir. It is possible to continue pumping the recovered CO_2 into the same reservoir and, after the completion of oil-recovery operations, close the site, leaving the CO_2 in the reservoir (NETL, 2008). Currently in the United States, CO_2 is recovered mainly from natural reservoirs for use in EOR operations (Kuuskraa, 2006).

A conservative estimate is that higher oil prices could increase U.S. petroleum production using EOR by between 1 million and 1.5 million bbl/d and use between 350,000 and 750,000 tons per day of CO_2.[4] This amount of CO_2 represents the plant-site production associated with an FT CTL industry producing approximately 500,000 to more than 1 million bbl/d, assuming that the plants have similar characteristics to the one analyzed in Chapter Five. With appropriate operation of projects, much of the CO_2 could be permanently stored.[5]

Geologic Storage

Geologic storage refers to technical approaches being developed and demonstrated worldwide that are directed at the long-term storage of CO_2 in various types of geological formations, such as deep saline formations. In geologic storage, CO_2 is injected at high pressure into appropriate formations. Three ongoing large-scale tests of geologic storage worldwide seek to store CO_2 while gaining critical knowledge to be applied elsewhere, and others are planned (IPCC, 2005; NETL, 2007b). One, in Weyburn, Saskatchewan, uses CO_2 delivered via pipeline from a coal-gasification facility in North Dakota for EOR. Recently, the Weyburn test has increased its injection rate of CO_2 from an initial 1 million metric tons per year to more than 2 million metric tons per year. The Sleipner project, operated by Statoil in the North Sea, injects approximately 1 million metric tons/year of CO_2 separated from natural-gas processing into a saline formation. The In Salah project in Algeria injects CO_2 to increase natural-gas recovery. A common aspect of the three projects is detailed monitoring of the migration of the injected CO_2 over time so that risks associated with geologic storage can be better understood (IPCC, 2005). Each project has a final storage capacity of approximately 20 million metric tons, and all three projects currently are viewed as successes in the scientific and technical literature.

Technical barriers to geologic storage in appropriate geologic formations appear to be "manageable and surmountable," and storage "is likely to be safe, effective, and competitive with many other options on an economic basis" (MIT, 2007, p. 43). Furthermore, the existence of significant natural reservoirs of CO_2 currently providing the gas for EOR operations is evidence of the earth's ability to store CO_2 under appropriate conditions (Bartis, Camm, and Ortiz, forthcoming). Nevertheless, further large-scale testing is needed before geologic storage can be considered viable from both technical and policy perspectives. In addition to broader characterization of geologic formations to support storage, the development of an appropri-

---

[4]   Kuuskraa (2006). For CO_2 flooding, each ton of CO_2 yields between two and three barrels of petroleum.

[5]   NETL (2008). The assumptions underlying the analysis in NETL (2008) are that the world price of oil is $70/bbl and that CO_2 costs $45/ton. If oil prices are higher or if CO_2 costs (including infrastructure costs) are lower, the demand may be higher.

ate regulatory system and capacity to quantify and manage risks remains to be done (NETL, 2007b).

CHAPTER FOUR

Oil Sands and Synthetic Crude Oil

This chapter considers the use of oil sands to produce liquid fuels via the extraction of bitumen and conversion to SCO. The analysis includes a discussion of the North American oil-sand resource base, the technologies involved in extracting and upgrading bitumen to SCO, the CO_2 emissions associated with oil-sand operations, and options for capturing and storing plant-site emissions of CO_2. It also addresses limiting factors with respect to the expansion of oil-sand development in Canada.

Overview of the Resource

Oil sands are deposits of bitumen in sand or porous rock. Bitumen, a mixture of hydrocarbons that, at normal temperatures and pressures, is a solid or semisolid, tarlike substance,[1] is the material of interest in oil-sand extraction. Bitumen extraction is performed by mining or by in situ methods. Oil sands are defined by ranges of viscosity and density. Their exact properties differ from region to region, but they are generally characterized by a high density (low American Petroleum Institute [API] gravity) and high viscosity. Since bitumen does not flow under ambient conditions, it is more difficult to recover than conventional crude oil is and requires significant subsequent upgrading to become a substitute for conventional crude. The upgrading process centers on adding hydrogen to the bitumen at elevated temperatures and pressures, yielding SCO.[2] The SCO that is produced can be refined into various petroleum products, such as gasoline, diesel, and jet fuel, in existing refineries in much the same way that conventional crude oil is refined (Alberta Chamber of Resources, 2004; NEB, 2006; Speight, 2007).

Companies process bitumen into one of three products: SCO, dilbit, or synbit. SCO is the upgraded product that refineries can use as a substitute for crude oil. Some companies decide not to upgrade the bitumen and instead sell it to refineries that have the capability to upgrade and refine the bitumen. However, to transport the bitumen to the buyer, it is mixed with either SCO or a condensate, such as naphtha, to decrease its viscosity and allow the product to flow through pipelines. When mixed with SCO, the result is called *synbit*. If naphtha or another condensate is used, the product is called *dilbit* (Nakamura, 2007). The process flow for each of these products is shown in Figure 4.1.

---

[1] Hence the term *tar sands*, which is generally used to describe oil-sand deposits outside of Canada, most notably those in Venezuela. However, since *tar* refers to destructive distillation products and not naturally occurring hydrocarbons, this term is not technically correct.

[2] *SCO* is a term that can also be used to describe upgraded oil shale and certain types of coal-derived liquids. However, in this document, we use the term exclusively to refer to the product derived from oil sands.

Figure 4.1
Oil-Sand Products

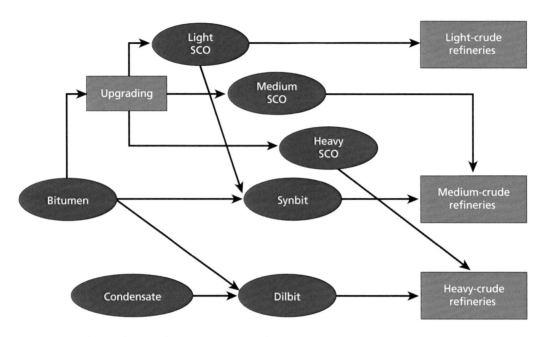

SOURCE: Based on Nakamura (2007, p. 56, Figure 4).
RAND *TR580-4.1*

North American Oil Sands

Resource Base

Canada has very large oil-sand reserves.[3] The remaining established reserves of bitumen in Canada are estimated to be 173 billion barrels (Alberta Energy Resources Conservation Board, 2008). This places Canada behind only Saudi Arabia in oil reserves. Canada's major oil-sand deposits reside in Alberta in three fields: Athabasca, Peace River, and Cold Lake. The quality of an oil-sand deposit can vary considerably depending on the thickness of the deposit (the *pay thickness*) and the saturation of bitumen in the deposit. Pay thickness varies from 19 to 100 feet, and saturation varies from 4.7 percent to 10.2 percent by mass in major deposits (Alberta Energy Resources Conservation Board, 2008).

U.S. resources of bitumen have not been heavily exploited and are not characterized as thoroughly as resources in Canada (USGS, 2006). Major deposits of bitumen (i.e., larger than 100 million barrels) in the United States can be found in Alabama, Alaska, California, Kentucky, New Mexico, Oklahoma, Texas, Utah, and Wyoming. The largest volume is in Utah, which has measured reserves of 8 billion to 12 billion bbl and total resources in place, including speculative ones, of 23 billion to 32 billion bbl. The total U.S. oil-sand resource is estimated at

---

[3]   In this document, we focus on North American resources, primarily those in Canada. This is due to (1) the magnitude of the Canadian reserves and (2) the relative maturity of the oil-sand industry in Canada. Venezuela also has vast oil-sand deposits north of the Orinoco River that rival the resources in Canada. However, these differ in character from Canadian oil sands and have not been developed for liquid-fuel production.

54 billion bbl in the form of bitumen, of which 22 billion are considered to be *measured*, and 32 billion are considered *speculative*.[4] These estimates of resources in place do not indicate how much of the resource may be recoverable or at what cost.

U.S. and Canadian oil-sand resources are significantly different in character,[5] and extraction techniques for the two will differ accordingly (BLM, 2007). The rugged terrain and specific geology of the Utah oil-sand deposit are likely to make extraction in the United States more challenging (U of U, 2007). Although U.S. oil-sand resources are significant, within the next few decades, they are not likely to add significant liquid-fuel capacity to U.S. production (Bartis, Camm, and Ortiz, forthcoming; DOE, 2007). Section 369 of the Energy Policy Act of 2005 (P.L. 109-58) directed the U.S. Department of Energy to make available for lease public lands for research and development and perform a programmatic environmental-impact statement regarding the production of liquid fuels from oil shale and oil sands (BLM, 2007). As of late 2007, no leases had been issued for the development of U.S. oil sands, and production to date is limited to two pilot-scale operations (U of U, 2007). Given anticipated limited production, as well as limited information on the costs of any production, we focus in this analysis on Canadian oil-sand extraction.

Production Projections

There have been a number of estimates of expected future production levels of SCO from Canadian oil sands. The Alberta Energy Resources Conservation Board expects production of Canadian oil sands to grow to 3.2 million bbl/d in 2017.[6] Similarly, the Canadian Energy Research Institute (CERI) estimates that, by the 2015 to 2020 time frame, production levels will be between 3 million and 4 million bbl/d.[7]

The National Energy Board (NEB) (2006) provided a broader range of production estimates: between 1.9 million and 4.4 million bbl/d in 2015. However, it considered the high-end estimate of 4.4 million bbl/d, based on all publicly announced projects as of June 2006 going online as originally scheduled, to be "beyond the limits of capacity growth" by 2015. The low-end estimate is based on sustained oil prices below $35/bbl. Since its base-case estimate of 3.0 million bbl/d by 2015 is based on an assumption that oil prices will be $50/bbl, it might be considered conservative.[8]

EIA also has made projections for Canadian oil-sand and bitumen production in 2025. In 2007, it projected that production levels in 2025 would be 3.2 million bbl/d in the reference case and 3.9 million bbl/d for the high-price case (EIA, 2007b, Tables G3 and G6). In 2008, in the context of higher prevailing oil prices, it projected 3.8 million bbl/d and 7.5 million bbl/d for its 2025 reference and high-oil-price cases, respectively (EIA, 2008b, Tables G3 and

---

[4] U of U (2007). Speculative resource estimates are highly uncertain because they are typically the result of extrapolating observations of surface resources below the surface (USGS, 2006).

[5] For example, U.S. oil sands are hydrocarbon wetted, whereas Canadian oil sands are water wetted.

[6] Alberta Energy Resources Conservation Board (2008). Of this, 1.9 million bbl/d are expected to be upgraded to SCO, and 1.2 million bbl/d are expected to be marketed as dilbit or synbit, which will be upgraded and converted to petroleum products at appropriately equipped refineries throughout North America, taking into account expected changes in refinery capacity and capability to upgrade bitumen.

[7] Timilsina, LeBlanc, and Walden (2005, p. ix). CERI also expects $100 billion in investments by 2020 in Canadian oil-sand projects.

[8] The base case assumes capital expenditures of C$94 billion during this time.

G6). These projections emphasize that future production levels are uncertain and dependent on world oil prices, as well as environmental impacts, technological progress, and other factors that will be discussed in more detail later.

If world oil prices remained significantly higher than the production cost of SCO, there would be large economic benefits from significantly expanding oil-sand production, and environmental considerations would weigh more heavily on capacity expansion—though the large economic returns would also make possible some significant efforts to mitigate environmental risks. Sustained oil prices around or below the production cost of SCO would dampen investment and production levels, and more detailed scrutiny of the economics of specific projects would be needed to evaluate the risk of investment.

Methods of Extracting and Upgrading Oil Sands

As noted, oil sand–extraction techniques fall into one of two general categories: mining or in situ. Mining is the method of choice if the deposit is covered by no more than ~250–330 feet of overburden, while in situ methods are preferred for deeper deposits. More than 60 percent of production of Canadian oil sands to date has been through mining, and mining will continue to be the dominant technique to recover bitumen in the next decade, with the share of in situ production expected to increase only slightly from 41 percent in 2007 to 45 percent in 2017 (Figure 4.2).[9] However, most of the oil-sand reserves (~80 percent) reside in areas where only in situ methods are feasible,[10] and minable oil-sand deposits are found only in the Athabasca deposit (Alberta Energy Resources Conservation Board, 2008). There are multiple in situ methods, including a number of promising techniques still in the development phase, but here we focus on the most common methods currently in use: steam-assisted gravity drainage (SAGD) and cyclic steam stimulation (CSS), both of which involve injecting steam into the ground to separate the bitumen from the sand and decrease its viscosity (Alberta Chamber of Resources, 2004; NEB, 2006).

Mining
Mining of oil-sand deposits is akin to strip mining. The process involves the use of large shovels, capable of moving up to 100 tons/scoop, to place the oil sands into large trucks, which often have capacities on the scale of 400 tons. To separate the bitumen from the inorganic matrix (primarily sand and clays), the mined material is brought to a mine site, where it is crushed, mixed with water, and pipelined as a slurry to a nearby extraction facility.[11] At the extraction facility, the slurry enters vessels designed to separate solids based on the differences between their densities (gravity separation) or surface properties (froth flotation).[12] The bitumen froth would then be sent to an upgrading facility, where it would be further processed into marketable products.

---

[9] Other estimates show the 60-40 split between in situ and mining continuing in this period.

[10] Of the 173 billion barrels of Canadian crude-bitumen reserves, it is estimated that 31 billion of those barrels may be recovered via surface mining and that 142 billion will be recovered via in situ extraction.

[11] This hydrotransport method is relatively new but gaining wide acceptance, replacing or supplanting earlier methods in which dry oil sands are transported (via truck and conveyor belts) directly to the extraction plant.

[12] Separation is enhanced by the addition of chemicals to the slurry that facilitate gravity separation or flotation.

Figure 4.2
Canadian Bitumen Production: Past and Future Projected

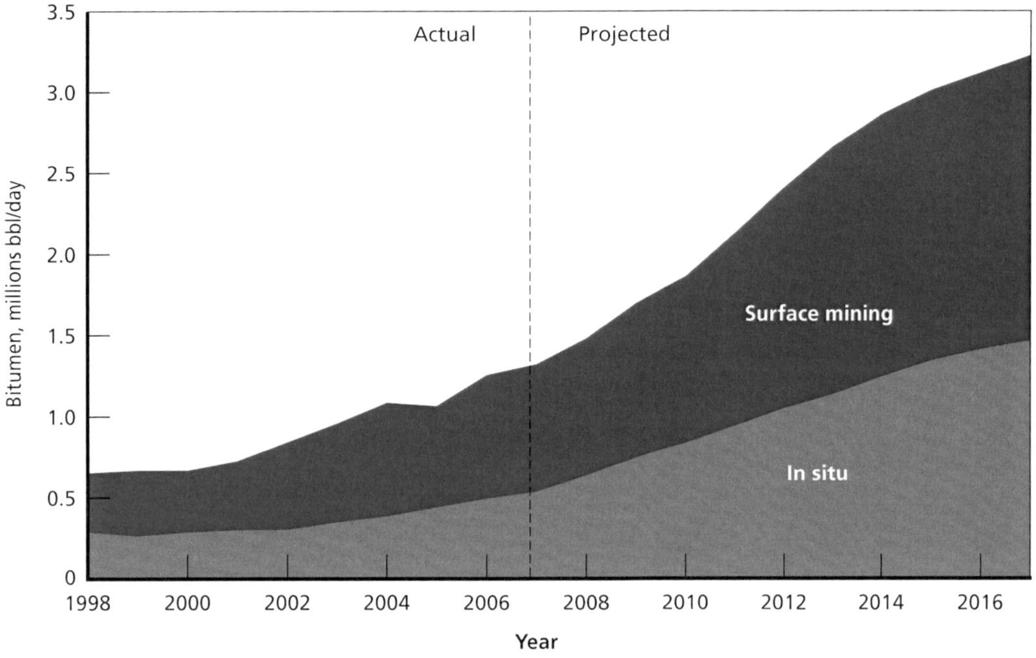

SOURCE: Based on information from Alberta Energy Resources Conservation Board (2008, p. 2-20, Figure 2.12).
RAND *TR580-4.2*

About 90 percent of the bitumen can be recovered during extraction. The sands and unrecovered bitumen that collect at the bottom of the separation vessel are known as *tailings*. These tailings are mixed with large quantities of residual water from the separation process and subsequently piped to tailing ponds. Here, the sand and clay are allowed to settle out so that water may be reused in the oil sand–separation process. However, there are limits to how much reuse is possible. Additionally, there are significant potential challenges associated with managing the tailing ponds (see section on "Environmental Impacts and Water Resources") (Speight, 2007).

Steam-Assisted Gravity Drainage

SAGD is an in situ method in which two closely spaced horizontal wells are drilled into the bitumen deposit. Steam is continuously injected into the top well at low pressures, and good vertical permeability within the deposit is required. The steam-to-oil ratio (SOR) indicates how many barrels of water (in the form of steam) are required to produce a barrel of oil and is the primary metric used to assess a SAGD site. A high-quality SAGD reservoir has an SOR of ~2.5, but this value can vary widely even within a given site or for a particular operation.[13] At present, steam in SAGD operations is generally created by burning natural gas. The steam decreases the bitumen's viscosity and carries it to the lower production well. The bitumen is then pumped to the surface. SAGD produces a recovery rate of 40 to 70 percent of resources in

---

[13] An SOR of 2.5 is also the assumed value in the MIT model we use in this analysis, as described later (see Lacombe and Parsons, 2007).

place (Alberta Chamber of Resources, 2004; NEB, 2004; Griffiths, Woynillowicz, and Taylor, 2006).

Cyclic Steam Stimulation

CSS (also sometimes referred to as *huff and puff*) is similar to SAGD in that it uses steam to decrease the viscosity of the bitumen. However, the technique uses a single vertical injection and production well, which is run cyclically instead of continuously as in SAGD. CSS is run in three phases: steam injection, soak, and production. In the first phase, steam is injected into the well at high pressures. The steam is then allowed to soak into the deposit and decrease the bitumen's viscosity. Finally, the bitumen is pumped out of the same well. This process is continually repeated. The recovery rate is 20 to 35 percent, and the SOR tends to be higher than for SAGD, although CSS does not require the higher-quality steam and heat/unit volume that SAGD does. CSS also requires a shale cap layer to help maintain pressure during the soak phase. Most of the CSS facilities in Alberta are in the Cold Lake area, which has a shale cap layer in some places. However, the majority of Canadian bitumen deposits are in the Athabasca field, which does not have a shale layer, making CSS infeasible (Alberta Chamber of Resources, 2004; NEB, 2004; Speight, 2007).

Upgrading

Before bitumen can be used in existing refineries designed for conventional petroleum products, the physical and chemical properties must be altered by upgrading to SCO.[14] The specific primary and secondary upgrading processes selected depend on the properties of the bitumen feed and the desired properties of the upgraded SCO product.[15] Feed separation may be performed prior to upgrading to separate out the heavy components, known as *bottoms*, which then undergo primary upgrading; the lighter hydrocarbons may directly undergo secondary upgrading (see Figure 4.3). Primary upgrading produces lighter hydrocarbons and may or

Figure 4.3
Upgrading Flowchart

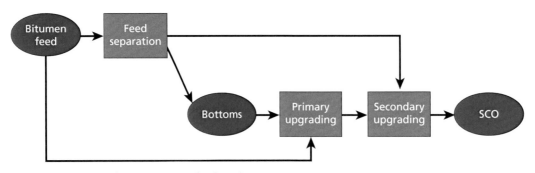

SOURCE: Based on information in U of U (2007).
RAND *TR580-4.3*

---

[14] Note that inclusion of upgrading is sometimes denoted by calling a project *integrated*. For example, an *integrated mining project* signifies that upgrading is part of the project. The output is thus a more valuable product than that from a project that is not upgraded. While the properties of SCO are similar to conventional crude oil, it is considered synthetic because it has been altered from the naturally occurring state (bitumen).

[15] *Enhanced upgrading* refers to novel, emerging upgrading processes, which will not be discussed herein.

may not utilize a catalyst, and secondary upgrading removes impurities from the feed and is a catalytic process. Primary upgrading in Canadian oil-sand operations commonly consists of thermal cracking via coking, which does not involve a catalyst.[16] This is generally followed by secondary upgrading via hydrotreating, a catalytic process done in a hydrogen-rich environment. The upgrading plants usually use natural gas for both heat and hydrogen production, making this a natural gas–intensive process. A coke that is very similar to petroleum coke is a by-product of primary upgrading. The coke can be used as a fuel for producing electricity or process heat (U of U, 2007). Without CCS, this use of the coke would result in higher CO_2 intensity than that of natural gas, although it might be less expensive overall even with regulatory requirements to manage CO_2.

Although SCO is often equated to West Texas Intermediate (WTI) in its quality, some producers upgrade the SCO to sell at a premium while others produce a lower-grade product that is less expensive than WTI. This difference is related primarily to the quality of the feedstock as well as the upgrading process used to produce the SCO. The sulfur content—whether it is sour or sweet (high- or low-sulfur, respectively)—also affects the SCO value.

Future Oil-Sand Technologies

The consumption of natural gas could significantly influence SCO-production costs, especially for in situ techniques. The oil-sand industry is looking at a number of new technologies to reduce or eliminate its dependence on natural gas (U of U, 2007). One method, vaporized extraction (VAPEX), is similar to SAGD in that it would be a continuous process with horizontal wells. However, instead of using steam, a solvent, such as ethane, propane, or butane, would be injected into the top well to decrease the viscosity of the bitumen. By one estimate, VAPEX would have a capital cost 25 percent less than SAGD and an operating cost of 50 percent less (NEB, 2004). The implementation of this method seems several years away, although, according to the National Petroleum Council (NPC), it could come online in 2010 (NPC, 2007, p. 47).

In situ combustion methods are also being explored. One such method is toe-to-heel air injection (THAI) (U of U, 2007). THAI uses two wells: one vertical injection well and a horizontal production well. The impetus for this method is that it could be used for deposits that are lower quality, thinner, and deeper. Air is injected into the vertical well, igniting the deposit. The heat decreases the viscosity of the bitumen, which then flows into the horizontal production well. As with VAPEX, NPC (2007, p. 47) estimated that THAI could be online in the 2010 time frame.

Others in the oil-sand industry are looking into gasifying bitumen or residual products from SCO production, such as asphaltenes, to produce a low- or medium-Btu fuel gas (the active components of which are CO and H_2) that can be burned to meet energy requirements. Gasification technologies appropriate to this application are well developed and used in other industries. Bitumen has about twice the amount of carbon/Btu that natural gas does, resulting in higher CO_2 emissions unless measures are taken to capture and store CO_2.

Using nuclear reactors to provide steam, electricity, or hydrogen for use in oil-sand projects would reduce CO_2 emissions in the extraction and upgrading of bitumen. NPC (2007, p. 48) estimated that producing a nuclear-power plant fit for the purpose would likely not occur until

---

[16] Hydroconversion is an alternate catalytic method of primary upgrading. It is more expensive than coking, but it also produces a greater yield of SCO. With coking, 30–40 percent of the bitumen forms coke.

2020–2030. However, Alberta Energy has since announced plans to build a 2,200-megawatt (MW) nuclear facility in the Peace River area as early as 2017.[17] As this would be the first nuclear plant to be built in the province, legal, regulatory, and public-opinion issues will need to be addressed prior to its realization.

The other primary extraction methods under consideration are relatively immature in their development. One method would use microbes to help recover bitumen. Another method includes the use of electromagnetic heating to decrease the viscosity of the bitumen, while CAPRI™ would use a catalyst in conjunction with the THAI method (the catalytic method was developed in part by the Petroleum Recovery Institute [PRI], hence the name). None is close to being implemented. Future technological advances could significantly alter the oil-sand industry if they become commercially viable and are implemented. However, the time frame for them to be introduced into commercial practice is difficult to predict.

Potential Constraints on Oil-Sand Production

Environmental Impacts and Water Resources

Extracting oil sands on a large scale has considerable environmental impacts, and concern for these has risen with the growth of the industry (NEB, 2006). The oil sands lie within the boreal forest, which extends across Canada to northern Europe, Russia, and Alaska and is the second-largest forest system in the world. According to the Western Boreal Conservation Initiative,

> The boreal forest is unique for its geographic diversity and extent, and the abundance of its wildlife. [It] is important to more than half of all the country's migratory birds, providing critical nesting habitat during the breeding season [and plays] a critical role as a carbon reservoir for the world. With its numerous bogs, fens, lakes, rivers, and wetlands, the boreal region is also a major source of water and wetland habitat. (Environment Canada, 2007b)

Although oil sands have been exploited for decades and companies have tried to reclaim the lands they disturb, it remains uncertain whether land-reclamation methods currently employed will be successful (NEB, 2006). According to their most recent sustainability reports, as of 2006, Suncor and Syncrude had reclaimed less than 10 percent and just over 20 percent of land disturbed by their operations, respectively (Suncor, 2007a; Syncrude, 2006). These numbers are consistent with those reported in 2004 by NEB (2004). The Government of Alberta recently issued the first official reclamation certificate, so most of the reclamation to date has yet to be validated.[18] In addition to the footprint of the extraction sites themselves, extensive roads and pipelines are required to move equipment and labor to, and oil-sand products from, remote sites in often-pristine environments (Bordetsky, 2007).

---

[17] Ontario nuclear operator Bruce Power, majority owned by TransCanada, has since acquired Alberta Energy's assets. In December 2007, Bruce Power Alberta announced that it would "now begin the process toward launching a full environmental assessment of the Peace Country site for potential nuclear generation" (Platts Global Power Report, 2007). It is unclear when (or whether) the actual construction will go forward.

[18] This certificate was issued to Syncrude for a 104-hectare parcel of land known as Gateway Hill. See Government of Alberta (2008a).

Both mining and in situ extraction methods use a significant amount of water relative to the extraction of conventional crude oil. For the mining operations, the Athabasca River is the primary source of water, and oil-sand projects are by far the largest user of the Athabasca, at more than twice that of the city of Calgary (Woynillowicz, Severson-Baker, and Raynolds, 2005; Griffiths, Woynillowicz, and Taylor, 2006). Production of one bbl of SCO by mining requires between 2 and 4.5 bbl of water. As of June 2006, oil-sand projects had licenses to withdraw 2.3 billion bbl of water/year from the Athabasca, most of which ends up in tailing ponds. If all of the existing, approved, and planned projects were realized, this would result in licenses for about 4.3 billion bbl/year (NEB, 2006). The government of Alberta has addressed the issue with legislation limiting the maximum allowed total water withdrawal for all existing, approved, and planned uses, at most an annual withdrawal of 6.2 percent of the total annual volume of the minimum flow year on record (Alberta Environment, 2004). However, without an impact study, it is difficult to understand how this would affect the river basin. The Pembina Institute has expressed concern for the aquatic ecosystem of the river, as well as wetlands and peatlands across the region. In particular, the seasonal variability in the flow of the Athabasca River could be problematic; in winter months, the flow can drop to less than 15 percent of its average peak flow in July (Griffiths, Woynillowicz, and Taylor, 2006). NEB (2006) concluded, "the Athabasca River does not have sufficient flows to support the needs of all planned oil sands mining operations. Adequate river flows are necessary to ensure the ecological sustainability of the Athabasca River." The Canadian government is increasingly attempting to address the issue, including setting specific limits on water removal from the Athabasca River under a new water-management framework (Alberta Environment, undated).

As part of the extraction process for surface-mined oil sands, large tailing ponds are used to facilitate the separation of water from a slurry of oil, sand, silt, and clay. NEB (2006) noted, "almost all of the water withdrawn for oil sands (mining) operations ends up in tailing ponds." As of May 2006, tailing ponds already covered an area of more than 30 square miles (Griffiths, Woynillowicz, and Taylor, 2006), and the volume of ponds produced by Suncor and Syncrude alone are expected to exceed 6.3 billion bbl by 2020 (NEB, 2004). These extensive ponds contain hazardous contaminants, such as mercury and napthenic acids, that could migrate to groundwater or leak into surrounding surface water and soil and that pose a danger to migratory birds. Additionally, the ponds are significant sources of methane emissions (NEB, 2004; Griffiths, Woynillowicz, and Taylor, 2006). Long-term management of tailing ponds "is one of the main challenges for the oil sands mining industry" (NEB, 2004), and the ponds could become major public liabilities should the companies not be able to cover the clean-up costs.

As mentioned, most bitumen is accessible only by in situ methods. These can be less environmentally disruptive than surface mining in several respects. For example, in situ methods do not result in the extensive topographical changes that accompany surface-mining operations, and they do not lead to the creation of tailing ponds.[19] However, surface-based drilling and supporting infrastructure (including roads, power lines, and pipelines) cause displacement of other land uses and of preexisting flora and fauna at the site. While in situ production requires a significant amount of water, 90 to 95 percent can be recycled so that only 0.2 bbl of net additional groundwater/bbl of SCO is required, an order of magnitude less than the

---

[19] Compaction and surface damage can occur as a result of underground removal of bitumen.

net volume required for mining operations.[20] Demand for fresh water for in situ operations is expected to reach 82 million bbl/year by 2015, up from 32 million bbl/year in 2004, compared to the 2.3 billion bbl of water licensed for use by Athabasca mining operations in 2006 (NEB, 2006). We note that water for subsequent upgrading is also required. The main environmental concern with using water for in situ operations is that the water is generally drawn from aquifers (fresh or saline) due to the location of operations. The potential long-term effects of this practice are unclear but could include removal of fresh water from the watershed, drawdown and depressurization of freshwater aquifers, changes in groundwater levels, changes in underground water storage or flow due to voidage zones left by bitumen removal, and mobilization of naturally occurring heavy metals, especially arsenic (Griffiths, Woynillowicz, and Taylor, 2006). The government of Alberta's environmental department, Alberta Environment, is working to reduce or eliminate freshwater use for in situ projects, and the trend is to use more brackish or saline water. However, both saline and recycled water must be treated, resulting in sludge or solid waste that must be landfilled, requiring additional energy inputs (NEB, 2006; Griffiths, Woynillowicz, and Taylor, 2006).

Quantifying the importance of these issues to oil-sand production, either in terms of the extent of the environmental impact or in terms of the additional cost that might be associated with SCO production as a result of higher water prices or stricter environmental regulations, is beyond the scope of this report. NEB (2004) noted, "cumulative environmental effects of development are beginning to be considered in a coordinated manner. Oil sands developers are taking advantage of new opportunities and technologies as well as synergies in their operations to improve environmental performance." The degree of progress in resolving environmental issues will almost certainly affect the general desirability of using this resource on a large scale. On the other hand, at crude-oil prices substantially higher than the production price of SCO, such as the prices observed in 2008, the oil-sand industry would be able to finance substantial efforts to mitigate environmental risks.

Natural-Gas Prices

Extracting oil sands relies heavily on the use of natural gas. A comparison of spot prices of natural gas and crude oil shows that they have comparable levels of volatility (Figure 4.4). In addition, the amount of natural gas used by oil-sand projects has been increasing dramatically (see Figure 4.5). By 2015, more than 2 billion cubic feet/day (bcf/d) will be required. This is compared to Canada's natural-gas production of 17.8 bcf/d in 2004.

To reduce natural-gas requirements, companies are considering creating fuel gas from residual products from upgrading bitumen. OPTI Canada Ltd. will use this process in its Long Lake SAGD project, currently being built, which is expected to dramatically reduce natural-gas requirements (NEB, 2006; OPTI Canada, 2007). Although this increases up-front capital expenditures, it is expected to have a lower operating cost than other oil-sand projects. However, it will also increase CO_2 production in the absence of CO_2 controls.

Projections of future oil-sand costs could change due to a technical advance that addresses use of water as a solvent and the cost and availability of natural gas. As discussed, oil-sand companies are researching alternatives to using water as a solvent for extracting bitumen. The

---

[20] Note that, while in situ water recycling reduces the total water use relative to that implied by the earlier-noted SOR of 2.5, it does not reduce the amount of energy required to convert 2.5 bbl water/bbl SCO to steam for the extraction process.

Figure 4.4
Natural-Gas Prices Compared to Oil Prices

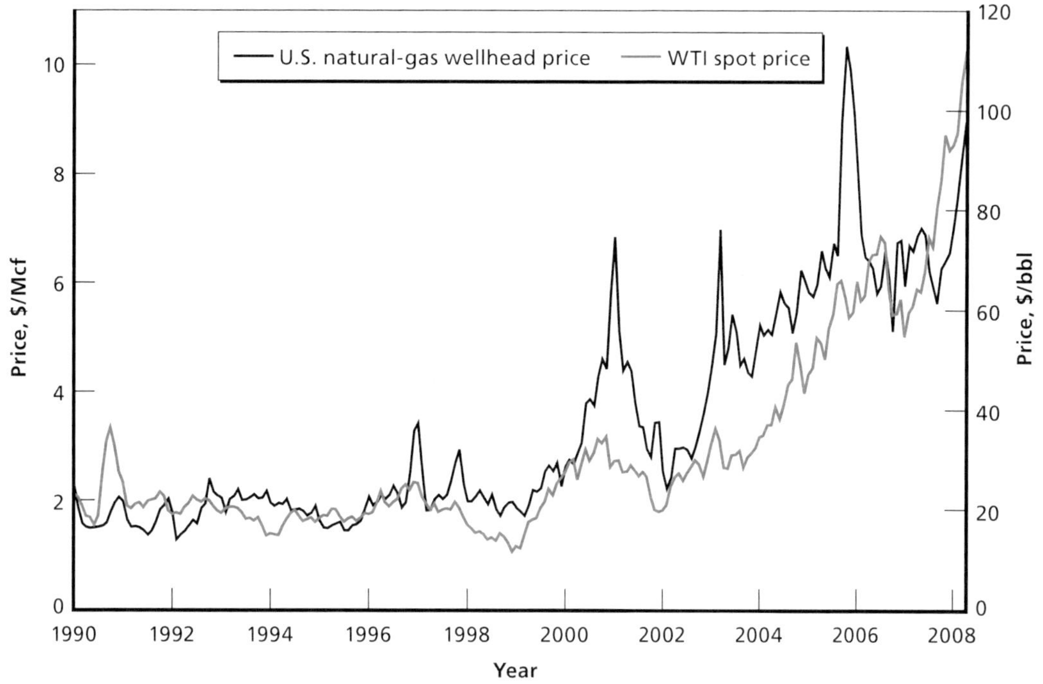

SOURCES: EIA (2008d, 2008c).
RAND TR580-4.4

two alternatives, VAPEX and THAI, are both in the early stages of research and appear to be years away from development. Publicly available information is not sufficient to develop cost estimates for these methods. In addition, the environmental consequences of injecting such solvents are unknown. There are too many uncertainties with these techniques to provide cost estimates.

Nuclear power could be used to produce electricity, steam, and hydrogen for oil-sand projects. However, in addition to concerns about radioactive-waste management and proliferation, there may be limitations on the use of nuclear power in the oil-sand industry. Oil-sand projects are generally dispersed, whereas nuclear plants generally provide a large amount of power at a single site. Piping steam over great distances would not be practical,[21] and electricity transmission would require significant infrastructure investments to reach many small, often remote oil-sand sites. H_2 production via electrolysis today is expensive, and, again, there is no existing infrastructure for moving large amounts of H_2 to remote oil-sand sites (NPC, 2007). At present, there is insufficient information to provide cost estimates if nuclear power were used in oil-sand projects. Additionally, if other methods, such as VAPEX or THAI, are introduced to replace the use of steam, these methods would compete with nuclear power. This is particularly true given that it will likely take at least a decade if not longer to build a nuclear-power plant, allowing significant time for one of these other methods to develop.

[21] The Alberta Chamber of Resources (2004, p. 54) stated, "more work is required on an economically attractive scale, as SAGD operations of greater than 100,000 barrels daily are normally spread out over an area not suitable for steam distribution from a single source."

Figure 4.5
Natural-Gas Consumption for Oil-Sand Production: Past and Future Projected

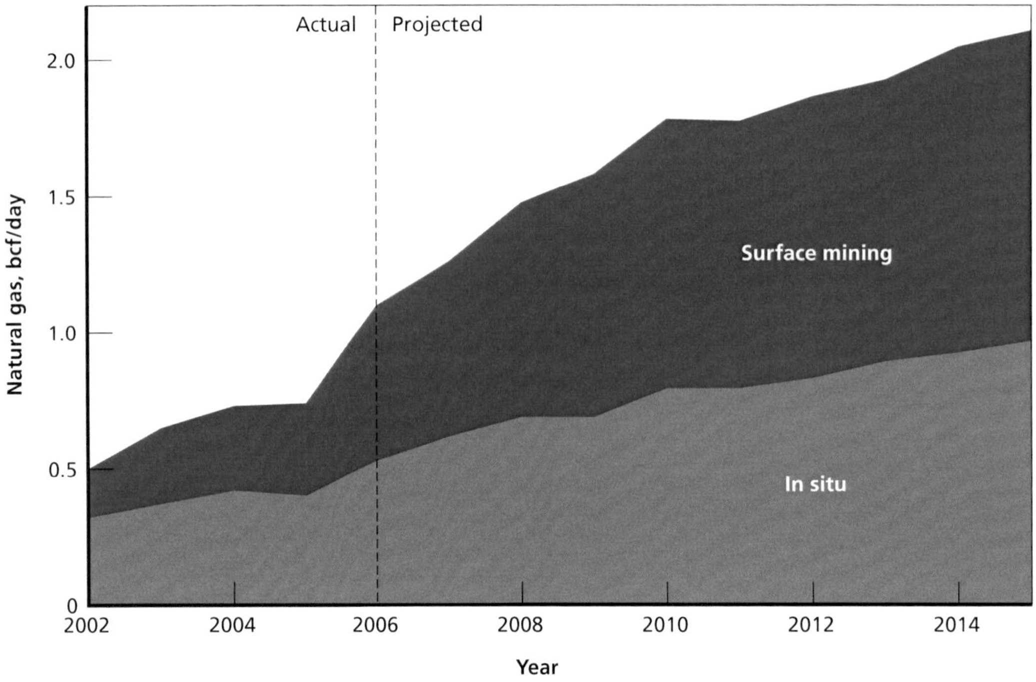

SOURCE: NEB (2006, p. 16, Figure 3.7).
RAND *TR580-4.5*

Other Market Constraints

The production of oil sands occurs within the context of a broader energy network, a limited labor market, and an evolving regulatory framework. For example, in addition to prices for natural gas, the physical availability of gas in terms of regional production and transportation bottlenecks will depend on the broader energy infrastructure in the oil-sand region and throughout Canada. Refining capacity and oil pipelines could also constrain the production of SCO on the output side. Additionally, the labor available in Alberta to build and run oil-sand projects is already insufficient. Companies have begun to fly workers in for periods of time and to offer significant bonuses, and labor expenditures could significantly raise the costs of future projects, or at least delay them, resulting in lower-than-anticipated production levels. Additionally, the oil-sand regions are faced with housing shortages and stress on public infrastructure and services (NEB, 2004, 2006). The rapid evolution of the political and regulatory environment makes it unclear what impact these factors will have. It is conceivable that regulatory and socioeconomic issues could significantly constrain the development of oil-sand production. While we do not attempt to quantify these impacts, and while "[s]takeholders have demonstrated a strong dedication to preserving the social well-being of communities" (NEB 2004), we note their potential significance to the profitability as well as the general feasibility and public acceptance of expanded oil-sand operations.

Carbon-Dioxide Production, Capture, and Storage

Producing SCO requires more energy input than does producing conventional crude oil. This implies higher life-cycle CO_2 emissions for SCO than for conventional crude and, in the face of CO_2 regulation, this could influence the relative economics of the two products.

Baseline Carbon-Dioxide Emissions from Oil-Sand Production

CO_2 emissions are created across all stages of bitumen extraction and upgrading to SCO, including from point sources that, in principle, are amenable to application of CCS. The total life-cycle CO_2 emissions for SCO are about 20 percent higher than low-sulfur, light crude oils.[22] Ranges of CO_2e[23] intensities used in this analysis for the individual steps in both mining and in situ SCO production are listed in the top half of Table 4.1 and are taken from a recent report by the Pembina Institute.[24] The ranges of total production emissions are 0.073–0.11 CO_2e/bbl SCO for mining and extracting plus upgrading and 0.094–0.13 tons CO_2e/bbl SCO for in situ plus upgrading. Pembina's numbers are consistent with values assumed in version 1.8a of the Greenhouse Gases, Regulated Emissions, and Energy Use in Transportation (GREET) model (Argonne National Laboratory, 1999), which are 0.099 and 0.12 tons CO_2e/bbl SCO for mining and in situ, respectively.[25] Using Suncor data for 2006, we calculate that the average emissions from its operations, which include both mining and in situ recovery operations and upgrading, are approximately 0.087 CO_2e/bbl SCO.[26] By Syncrude's own calculations, the average emissions from its combined mining and in situ operations are 0.12 tons CO_2e/bbl SCO.[27] Pembina's reported ranges are consistent with these industry values.

We account for the emissions associated with transporting SCO to the United States, and for the analogous transport emissions for crude oil, using values taken from GREET.[28] In this analysis, we assume CO_2 emissions from the refining, distribution, and use of a barrel of SCO to be the same as the analogous emissions from a barrel of crude oil. Specifically, we assume that a barrel of crude oil and a barrel of SCO each contains 0.13 tons of carbon, so the CO_2

---

[22] In practice, conventional crude comes from a variety of sources with a range of GHG-emission intensities; some of the heavy oils are more similar to oil sands with respect to total life-cycle GHG emissions.

[23] CO_2e refers to CO_2 equivalents and includes CO_2 as well as other GHGs, such as methane (CH_4) and nitrous oxide (N_2O), scaled by the heating impacts of the gases relative to CO_2.

[24] McCulloch, Raynolds, and Wong (2006). The Pembina Institute, an environmental organization focused on sustainable energy, is widely accepted as a reliable source for these figures. For example, NEB's Canadian–oil-sand report (2006) cited GHG-intensity values from an earlier Pembina report (Woynillowicz, Severson-Baker, and Raynolds, 2005). We also note that conversations with individuals in academia and industry have confirmed the general credibility of Pembina's figures and its analytic methods.

[25] These totals include CH_4, N_2O, and CO_2 intensities from GREET, scaled by IPCC global-warming–potential (GWP) values of 25, 198 and 1, respectively. Extraction, upgrading, and transport to U.S. facilities are all included.

[26] Environment Canada (2007a); Suncor (2007b, 2007c). This calculation is based on dividing total annual emissions by total production, of which 3 percent is bitumen, not SCO, so the value would be slightly higher/bbl SCO when including upgrading of this 3 percent of product.

[27] Syncrude (2006). Our calculations using Syncrude's reported production levels (Syncrude, 2006, p. 16) and the slightly higher Environment Canada (2007a) GHG-intensity number yield the same value to two significant figures.

[28] Emissions for recovery and transport of crude oil in the GREET model are 0.042 tons CO_2e/bbl and have been included in this analysis for the life-cycle emissions of conventional petroleum. Again, CH_4, N_2O, and CO_2 values from recovery and transport in the GREET model are used.

Table 4.1
Oil-Sand Emissions: Production of SCO and Life Cycle with Carbon Capture and Storage

Process	CO_2 Intensity (ton CO_2e/bbl SCO)	
	Low	High
Extraction and production emissions without CCS		
Mining and extraction		
Point (utility heaters, power)	0.015	0.022
Nonpoint (fleet, mine face, tailing ponds)	0.010	0.014
Total	0.025	0.035
In situ (SAGD)		
Point (boilers and plant energy)	0.044	0.050
Nonpoint (fugitive, vehicle fleets, flaring)	0.0036	0.0045
Total	0.047	0.054
Upgrading		
Point: hydrogen production	0.013	0.037
Point: combustion sources (coker, boilers)	0.034	0.034
Nonpoint (fugitive)	0.00091	0.00091
Total	0.047	0.072
Total production emissions generated		
Mining and extraction + upgrading	0.073	0.11
In situ + upgrading	0.094	0.13
Emissions assuming CCS with capture from point sources		
Mining and extraction + upgrading with CCS + transport to United States		
Production emissions captured and stored (assuming capture from all point sources)	0.053	0.079
Production emissions released	0.020	0.028
Transport emissions	0.021	Same
Use-phase emissions	0.47	Same
Total life-cycle emissions released	0.51	0.51
In situ + upgrading with CCS + transport to United States		
Production emissions captured and stored (assuming capture from all point sources)	0.076	0.10
Production emissions released	0.018	0.023
Transport emissions	0.021	Same
Use-phase emissions	0.47	Same
Total life-cycle emissions released	0.50	0.51

SOURCES: Point and nonpoint emissions in production: Pembina Institute (McCulloch, Raynolds, and Wong, 2006); emission-capture fraction from point sources: 86 percent for hydrogen production, 85 percent for all others (e.g., heaters, power) from IPCC (2005, Chapter Three); transport and use-phase emissions: GREET v1.8a (Argonne National Laboratory, 1999).

emissions resulting from refining and end use (i.e., combustion of products, such as gasoline) is about 0.47 tons for both.[29]

Carbon-Dioxide Capture and Storage for Oil Sands

CO_2 emissions for SCO production as shown in Table 4.1 are low relative to the use phase of the fuel. A large fraction of these are from point sources and thus can be captured. This analysis assumes that, if CCS were applied at all, it would be applied to all of these point sources. However, given that the cost of capturing CO_2 from hydrogen production will likely be lower than for other point sources, CCS could be applied just to upgrading facilities and not to extraction facilities.

Assuming the ranges presented in Table 4.1 for production emissions captured and stored, and assuming a 60-40 surface mining–in situ industry composition, a 3 million bbl/d oil-sand industry in Canada would produce around 70 million to 95 million tons of CO_2/year. Central Alberta has a significant number of natural-gas and oil fields, and the Alberta Chamber of Resources (2004, p. 65) concluded that there is enough storage space for about 100 million tons of CO_2 annually for more than 300 years. This CO_2 production level is at the high end of what we estimate would be produced by a 3 million bbl/d oil-sand industry, so even for the significantly larger production estimates for 2025 (i.e., ~7 million bbl/d), if these estimates are correct, Alberta would have ample geologic storage potential for a significantly expanded oil-sand industry. EOR is a preferred use of CO_2, but the total EOR potential of Canada or Alberta is not known.

There have been a number of recent activities related to CCS in Canada. For example, the Integrated CO_2 Network (ICO$_2$N), an alliance of 15 leading industrial companies in Canada, including Suncor and Syncrude, formed a CCS initiative in 2007. The group aims to work with the Alberta and Canadian governments on developing a multi-industry, multi-province CCS system, including a CCS policy framework and infrastructure (see ICO$_2$N, undated). Also in 2007, the Alberta government itself created a Climate Change and Emissions Management Fund for fast-tracking funding to low-emission technologies (Government of Alberta, 2008b). In March 2007, the prime minister announced the formation of the Canada-Alberta ecoENERGY Carbon Capture and Storage Task Force "to recommend the best ways for Canada to implement the technology on a large scale" (Government of Alberta, 2007). In April 2008, the government announced its plan to reduce CO_2 emissions by 20 percent from current levels by 2020. Also in April, the government of Alberta announced an agreement with TransAlta Corporation and Alstom Canada to develop a large-scale CCS facility in Alberta (TransAlta, 2008). In July 2008, the Alberta government committed to allocating $2 billion to "encourage construction of Alberta's first large-scale CCS project" as part of a $4 billion "climate change action plan" (Government of Alberta, 2008c).

Unit Costs for Oil-Sand Production

In this section, we present estimates of current unit costs for oil-sand production and estimate a range of potential production costs of SCO in 2025. To understand current and future SCO

---

[29] Our calculation ignores the small energy and CO_2 flows associated with any nonpetroleum-energy use in refining and with final product distribution.

prices in the context of potential natural-gas and CO_2 prices, we use the following equation for the cost of producing SCO using either mining or in situ methods:

$$
\begin{aligned}
\text{production cost} \left(\$ / \text{bbl}\right) = {} & \text{annualized capital charge} \left(\$\right) / \text{annual production} \left(\text{bbl}\right) \\
& + \text{annualized nonfuel operating costs} \left(\$\right) / \text{annual production} \left(\text{bbl}\right) \\
& + \text{natural-gas feedstock intensity} \left(\text{Mcf/bbl}\right) \\
& \times \text{price of natural gas} \left(\$/\text{Mcf}\right) \\
& + CO_2 \text{ emitted} \left(\text{ton } CO_2/\text{bbl}\right) \times \text{cost of } CO_2 \text{ emissions} \left(\$/\text{ton } CO_2\right) \\
& + CO_2 \text{ captured} \left(\text{ton } CO_2/\text{bbl}\right) \times \text{cost of CCS} \left(\$/\text{ton } CO_2\right).
\end{aligned}
$$

In this analysis, the first two terms in the production-cost equation are set within a fixed range of values for each of the two technologies. The third term provides a way to examine the sensitivity to natural-gas prices. The fourth term provides a way to account for the impact of regulations that would impose a cost on CO_2 emissions and is varied in this analysis. The final term is set within a fixed range of values based on our estimates of emissions and CCS costs.

Current Costs for Oil-Sand Production Without Carbon-Dioxide Management

Current cost estimates are derived from the technical and economic assumptions and resulting production costs reported by NEB (2004, 2006). Lacombe and Parsons (2007) developed a financial-analysis spreadsheet based on this information, which we use herein to determine the unit production costs of SCO in the absence of CO_2 management (i.e., to determine the values of the first three terms in the production-cost equation). The spreadsheet uses the standard discounted-cash-flow method to determine the cost of production of bitumen and SCO from each of the two dominant oil-sand technologies, integrated mining and upgrading and in situ SAGD with a separate stand-alone facility for upgrading.

The technical and economic assumptions that are used as inputs for the financial analysis are given in Tables 4.2a and 4.2b for the two technologies. The assumptions in column 2 of both Tables 4.2a and 4.2b are consistent with NEB and represent current costs as of 2006 (NEB, 2006; Lacombe and Parsons, 2007). The modified assumptions used in this analysis are summarized in column 3. The period of investment over which the internal rate of return (IRR) is calculated herein is the same as was assumed in the original 2006 NEB report, 45 or 42 years, depending on the technology.[30] The other assumptions in Lacombe and Parsons that we use are as follows: a capacity factor of 1.0, a refinery yield of 1 bbl SCO/bbl bitumen, and an exchange rate of Canadian to U.S. dollars of 0.85. All cost inputs and results in Table 4.2 are in real 2005 U.S. dollars (2005 US\$) and yield a levelized cost for products. The costs reported can be interpreted as the price obtained for SCO at Edmonton, which would produce a real, after-tax IRR of 10 percent. We have used the price of a barrel of SCO at Edmonton rather than making any assumptions regarding further shipping and refining costs.[31]

---

[30] The CTL cost model described here uses a 30-year period of investment.

[31] The figures are therefore comparable to those reported by NEB (2006) and other Canadian sources.

Table 4.2a
Economic and Technical Assumptions for Integrated Mining and Upgrading, Current and Future, Assuming No Costs for Carbon-Dioxide Management

Parameter	Current Value	Value or Range in 2025
IRR (%)	10	10
Investment period (years)	45	Same
Corporate tax (%)	32.1	Same
Royalty, minimum rate (%)	1	Same
Royalty, postamortization rate (%)	25	Same
Price of natural gas (2005 $/million Btu)	7	4–10
SCO production at full scale, by year 7 (bbl/day)	200,000	Same
Initial capital costs, equally spaced over initial 8 years of construction (2005 $ millions)	8,500	6,800–8,500
Recurring capital costs: years 5–6 at 100,000 bbl/d, years 7–45 at 200,000 bbl/d (2005 $/bbl SCO)	1.06	0.85–1.06
Natural-gas consumption (Mcf/bbl SCO)	0.75	Same
Non–natural gas cash operating costs (2005 $/bbl SCO)	10.20	8.16–10.20

SOURCES: Lacombe and Parsons (2007), NEB (2004, 2006).

Future Production Costs Without Carbon-Dioxide Management: Capital-Cost Uncertainties and Learning-Based Cost Declines

Even though SCO production from Canadian oil sands has been ongoing for decades, future production costs are uncertain. This is exemplified by the twofold increase that NEB (2004, 2006) estimated in production costs for 2004 to 2006. Moreover, according to NEB, oil-sand capital costs today are even higher than those indicated in its 2006 report due to rapidly escalating construction costs, including labor (NEB, 2008). The production-cost uncertainties center on the capital costs of future oil-sand projects (i.e., the first term in the production-cost equation). These increases are due to the cost-factor escalation for materials and labor used in capital projects that has been higher than anticipated and more rapid than general inflation. A number of other competing factors could either increase or decrease future production costs, and it is not clear which factors will dominate. For example, oil-sand developments can be highly site-specific, and future oil-sand development will increasingly target resources that are more difficult to extract than prior ones. Environmental remediation will also have associated costs. However, there are also opportunities to improve the efficiency of existing technologies and the potential for the development of new ones (Bergerson, 2007; Bergerson and Keith, 2006). We have attempted to capture some of this uncertainty in a simple way by including a fixed range of capital costs for both technologies in this analysis.

Production costs tend to decline as an industry matures. Historical experience curves can be used to project future supply costs based on expected future cumulative production (Lieberman, 1984; Argote and Epple, 1990). Historical data from the literature (Alberta Chamber of Resources, 2004; Alberta Energy Resources Conservation Board, 2008) indicate a significant decline in SCO production costs between 1985 and 2000. However, a simple trend analysis based on cumulative production is insufficient. First and most obvious, recent production costs

Table 4.2b
Economic and Technical Assumptions for Steam-Assisted Gravity Drainage and Upgrading, Current and Future, Assuming No Costs for Carbon-Dioxide Management

Parameter	Current Value	Value or Range in 2025
IRR (%)	10	Same
Investment period, SAGD (years)	42	Same
Investment period, stand-alone upgrader (years)	45	Same
Corporate tax (%)	32.1	Same
Royalty, minimum rate (%)	1	Same
Royalty, postamortization rate (%)	25	Same
Price of natural gas (2005 $/million Btu)	7	4–10
SAGD		
Bitumen production at full scale, by year 13 (bbl/day)	120,000	Same
Annual construction capital costs, initial 12 years (2005 $ millions/yr)	128	102–128
Annual recurring capital costs, exploitation in years 13–42 (2005 $ millions/yr)	25.50	20.40–25.50
Required diluent (% blend by volume)	33.3	Same
SOR (bbl water/bbl bitumen)	2.5	Same
Natural-gas consumption (Mcf/bbl water)	0.42	Same
Nongas cash operating costs (2005 $/bbl bitumen)	$2.98	$2.38–2.98
Stand-alone upgrader		
SCO production at full scale, by year 7 (bbl/day)	200,000	Same
Initial capital costs, equally spaced over initial 8 years of construction (2005 $ million)	$6,380	5,100–6,380
Recurring capital costs: years 5–6 at 100,000 bbl/d, years 7–45 at 200,000 bbl/d (2005 $/bbl SCO)	0.53	0.43–0.53
Natural-gas consumption (Mcf/bbl SCO)	0.47	Same
Nongas cash operating costs (2005 $/bbl SCO)	4.25	3.40–4.25

SOURCES: Lacombe and Parsons (2007); NEB (2004, 2006).

have been driven much higher than historical ones, as a result of economic and other constraints (NEB, 2004, 2006; Lacombe and Parsons, 2007). Additionally, most of the decrease in costs for integrated mining has come from improvements in mining practices that are mature; for in situ production, costs of production were flat for several years before the recent increases (Alberta Chamber of Resources, 2004).

The literature on learning curves suggests that a "typical" learning rate over the life of an industry implies a cost decline on the order of 20 percent with each doubling of cumulative production. There is a wide distribution around this value, depending on the technology considered and the study referenced. For example, a 1989 RAND study found, on average, a

24-percent decline in costs for every doubling of production capacity across a broad array of chemical-process technologies, but the declines for individual technologies ranged from 2 percent to 44 percent.[32] Taking the various contributors to oil-sand costs into account, including recent cost escalation, we use in this study a more conservative progress rate of approximately 10 percent for this more mature technology. Specifically, we assume that, since both oil-sand technologies are expected to approximately double twice in cumulative capacity by 2025, a 10-percent learning rate (i.e., 10-percent unit-cost decline with every doubling of cumulative production) would lead to a 20-percent decrease in these costs by 2025. We apply these savings to the current base-case, nongas cash operating costs and initial and recurring capital expenditures in Lacombe and Parsons' financial spreadsheet. This is the *optimistic* (low-cost) case we use to represent the impact of learning on oil sand–production costs. The *pessimistic* (high-cost) case assumes the same costs as NEB (2006) reported, reflecting the other uncertainties noted here, particularly those associated with increasingly difficult-to-extract oil-sand resources and inflation in labor and material costs. These ranges are summarized in the third columns of Tables 4.2a, 4.2b, and 4.3.

Arguments can be made for why future costs could be higher or lower than the range we report herein; there is not enough information and there are too many conflicting factors to definitively determine these values. Even so, it appears very likely that SCO-production costs, in the absence of a price on environmental externalities, such as CO_2 emissions, will be well under \$50/bbl, which means that SCO will be economically competitive with crude oil over a wide range of expected future price scenarios.

Cost Sensitivity to the Price of Natural Gas

In addition to technology-cost uncertainties, the intensity of natural-gas use in the production of oil sands can be highly variable. For example, the financial spreadsheet we use herein assumes a commonly cited SOR of 2.5 for a high-quality SAGD site (Lacombe and Parsons, 2007; NEB, 2006; Griffiths, Woynillowicz, and Taylor, 2006), but, in reality, this value can range over as much as an order of magnitude, even within a given site (Bergerson, 2007). In the discussion that follows, we examine the sensitivity of SCO cost to the price of natural gas. This can also be interpreted as providing a rough representation of the sensitivity of SCO cost to the intensity of natural-gas use.

Using the production-cost equation, we first examine the sensitivity of SCO-production costs to variations in the price of natural gas assuming that no costs are incurred for releasing CO_2 and that no CO_2 is captured (i.e., the fourth and fifth terms in the equation are 0).

Table 4.3
Unit Production Costs, Current and Future, Assuming No Costs for Carbon-Dioxide Management

Parameter	Current Costs	Range of Costs in 2025
Integrated mining and upgrading (2005 \$/bbl SCO)	34.00	27.70–33.50
SAGD and stand-alone upgrader (2005 \$/bbl SCO)	37.30	30.70–35.90

SOURCE: Lacombe and Parsons (2007) spreadsheet output based on assumptions in Table 4.2 and those described in the text about cost changes.

---

[32] Merrow (1989). Although the oil-sand example in this study was given as an example of failure to learn in pioneer plants, as noted, the oil-sand industry did experience significant cost decreases subsequent to 1989.

Figure 4.6 illustrates this production-cost sensitivity for SCO produced by both integrated-mining and in situ methods. We have covered a broad range of natural-gas prices including the range of prices predicted by EIA in 2007 for 2025. The orange and blue regions represent the uncertainty associated with the range of capital and operating costs for the two production techniques, integrated mining and SAGD, respectively. The vertical dashed line in Figure 4.6 shows the EIA 2007 reference-case value for the price of natural gas (for electric power) in 2025. The horizontal dashed line indicates EIA's 2007 reference-case price of imported low-sulfur light crude oil for the same year. Since the unit costs of both technologies rise with higher natural-gas prices, the break-even crude-oil price for both technologies is higher if natural-gas prices increase. However, on a Btu basis, the break-even oil price is not highly sensitive to the natural-gas price; each $1/mmBtu increase in the price of natural gas causes around $1/bbl increase in the break-even oil price, depending on the technology ($0.70/bbl and $1.50/bbl for mining and SAGD, respectively). Accordingly, the production cost for SCO from in situ processes is not expected to exceed the EIA base-case price of crude, even under the high capital-cost assumptions for SAGD, until the price of natural gas reaches nearly $20/million Btu (not shown in Figure 4.6, as this is not within the range of plausible natural-gas prices). In other words, under EIA's scenarios, natural-gas prices alone are not expected to constrain oil-sand development. We also note that, in all likelihood, future scenarios in which natural-gas prices will be relatively high are also those in which crude-oil

Figure 4.6
2025 Production Cost of Synthetic Crude Oil Versus the Price of Natural Gas, Assuming No Costs for Carbon-Dioxide Management

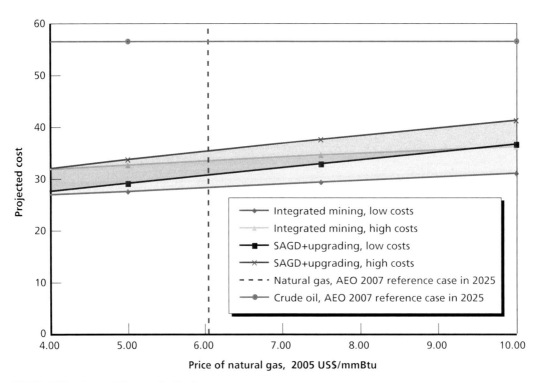

NOTE: AEO = *Annual Energy Outlook.*

prices will be high. For these reasons, we do not incorporate natural gas–price variability in our comparison of SCO and conventional-fuel costs in Chapter Five.

The higher average intensity of use of natural gas for SAGD causes its unit production cost to exceed that of integrated mining when natural gas costs more than \$3.00/mmBtu and \$3.50/mmBtu for the low- and high-cost cases, respectively. However, most oil-sand reserves are accessible only through in situ methods, such as SAGD. Therefore, in practice, this will be the marginal technology for significant expansion of SCO, even if natural-gas prices are around or above the EIA reference level.

Current Carbon Dioxide–Management Costs for Synthetic Crude Oil

A range of costs for production of SCO with CO_2 management in 2025 can be calculated assuming that CCS is employed at all point sources in the processes of extracting and upgrading bitumen into SCO for both mining and in situ methods. We can then determine a range of values for the fifth term in the production-cost equation. With these values, we can examine the sensitivity of SCO-production costs to variations in the cost of CO_2 emissions (the fourth term in the production-cost equation) both with and without CCS. We begin by estimating what the cost for CO_2 management for SCO operations would be today.

Estimates for the current costs of capture of CO_2 for all point sources are based on data compiled in the most recent IPCC *Special Report on Carbon Capture and Storage*.[33] For all point sources other than hydrogen production (i.e., utility heaters, power, boilers, combustion sources), the cost of controlling CO_2 from a new PC power plant was used;[34] new H_2-plant values were used for the H_2 point-source cost to control CO_2 (IPCC, 2005, Table 3.15). The CO_2 captured/bbl of production represents approximately an 85-percent reduction in CO_2 emissions from all point sources listed in Table 4.1.[35] In addition to the 15 percent of point-source emissions that are not captured, we assume that no non–point source emissions will be captured. All noncaptured emissions will instead result in an additional cost based on the cost of CO_2 emissions. The range of expected costs for CO_2 transport and storage are also taken from IPCC (2005, Table TS.9). Pipeline and infrastructure costs are proportional to the distance the CO_2 must be transported and to the size of the CO_2-generating facility. The costs of identifying, commissioning, decommissioning, and long-term monitoring of storage sites are highly uncertain.

Total current CCS costs are a sum of the following values from IPCC: (1) capture costs from heaters, power, and boilers: costs centered on the representative value for new PC plants; (2) capture costs from H_2 production: costs centered on the representative value for new H_2

---

[33] IPCC (2005). All IPCC values were converted from 2002 \$/metric ton (tonne) to 2005 \$/ton. To do this conversion, we used the ratio of the 2005 producer price index (PPI) for petroleum refineries (Bureau of Labor Statistics, undated) to the average of the values for this index in 2001, 2002, and 2003. This ratio yielded a multiplier of $1.92 \left(= 205.3 \big/ \left(\left(103.1 + 96.3 + 121.2 \right) \big/ 3 \right) \right)$. The rationale for using this particular PPI rather than a broader-based index is that (a) the refining sector is a reasonable proxy for cost adjustments relevant to investments in CO_2 capture and (b) there were significant real cost changes in the refining sector over the period in question, which a broader-based price index would not adequately capture.

[34] We use IPCC's cost/ton of CO_2 avoided rather than the cost/ton of CO_2 captured in recognition of the fact that CO_2 capture has an energy (and thus also a CO_2) penalty.

[35] Based on IPCC values, we assume capture of 86 percent of H_2-production emissions and 85 percent of all other point sources (e.g., heaters, boilers, power).

plants;[36] (3) transport costs: the range of values for facilities that can produce ~5 megatons (Mton) CO_2/yr;[37] (4) storage, including monitoring and verification: the entire range given by IPCC (2005, Table 8.2). These values have been converted to appropriate units, and we have assumed, as does IPCC, that 85- and 86-percent reductions in CO_2 are possible from heaters, power, and boilers and hydrogen production, respectively. These values are the current values presented in the second column of Table 4.4.

Development and Learning for Carbon-Dioxide Capture

Consistent with IPCC findings, we anticipate that technical progress may reduce CO_2-capture costs by 20–30 percent over approximately the next decade (IPCC, 2005). Accordingly, for SCO operations in 2025, as a conservative midpoint, we assume a 25-percent reduction from IPCC estimates in costs for CCS-operation capture and compression components for our capture and compression costs. We have then reduced capture costs from both (1) heaters, power, and boilers and (2) H_2 production by 25 percent to yield costs in the year 2025, in 2005 dollars. For transport and storage, we do not assume any cost improvement because pipeline transport is based on a mature technology and our estimates of storage costs assume that significant learning has taken place. The resulting cost parameters are summarized in the third column of Table 4.4.

Using these cost assumptions, total CCS costs/bbl SCO in 2025 for each of the two technologies can be calculated. These are given in column three of Table 4.5. These values are added to the unit production costs from Table 4.3 to determine the range of total estimated

Table 4.4
Carbon-Dioxide Reduction and Expected Carbon Capture and Storage Cost Parameters for Oil Sands in 2025

Parameter	Current Value	Range of Costs in 2025
CO_2 reduction: e.g., heaters, power, boilers[a]	85 percent	
CO_2 reduction: H_2 production[a]	86 percent	
CCS system-component cost parameters ($/ton CO_2)		
Capture, point source: e.g., heaters, power, boilers[a]	62.70–80.20	47.10–60.10
Capture, point source: H_2 production[a]	17.40–34.90	13.10–26.10
Transportation, 250-km pipeline or shipping[b]	3.70–5.80	Same
Storage, including monitoring and verification[b]	1.10–14.50	Same

NOTE: Values have been adjusted to 2005 US$ and, in the case of capture costs, have been reduced by 25 percent to account for technological learning by 2025 (i.e., the anticipated improvement by 2015 in IPCC, 2005).

[a] SOURCE: IPCC (2005, Chapter Three, Table 3.15).

[b] SOURCE: IPCC (2005, Chapter Eight, Figure 8.1 and Table 8.2).

---

[36] For both capture numbers, see representative values in IPCC (2005, Table 3.15). In both cases, the range we have used around this number (±$5/ton CO_2) is within the range of values shown here but is more appropriately weighted toward the representative values.

[37] See IPCC (2005, Figure 8.1). We have used the range of costs for a 250-km pipeline at 5 Mton CO_2/yr (5.5 Mton CO_2/yr), which is approximately 2002 US$2.10–3.30/tonne CO_2, and converted these to 2005 US$/ton CO_2. This is based on a 200,000 bbl/d facility and storage requirements at 0.07 ton CO_2/bbl SCO.

Table 4.5
Expected Total Carbon Capture and Storage Costs/Barrel of Synthetic Crude Oil in 2025

Process	Current Cost ($/bbl SCO)	Range of Costs in 2025 ($/bbl SCO)
Mining and extraction + upgrading		
Capture, all point sources[a]	3.41–5.94	2.55–4.45
Transport[b]	0.23–0.55	Same
Storage[b]	0.07–1.39	Same
Total CCS	3.71–7.88	2.85–6.39
In situ + upgrading		
Capture, all point sources[a]	5.23–8.27	3.92–6.20
Transport[b]	0.34–0.72	Same
Storage[b]	0.10–1.81	Same
Total CCS	5.67–10.80	4.36–8.73

[a] Sources: (1) IPCC (CO_2-reduction percent and capture costs: IPCC, 2005, Chapter 3, Table 3.15); (2) Pembina Institute (CO_2 sources and intensities: McCulloch, Raynolds, and Wong, 2006).

[b] Sources: (1) IPCC (transport and storage costs: IPCC, 2005, Chapter 8, Figure 8.1 and Table 8.2); (2) Pembina Institute (CO_2 sources and intensities: McCulloch, Raynolds, and Wong, 2006).

costs for production for both technologies, including CCS, at all point sources and assuming a zero cost for emitting CO_2, as shown in Table 4.6. These base values are then used for the CO_2 cost–sensitivity analysis in Chapter Six.

Overall, our analysis gives a range of production costs in 2025 based on modest technological improvements to today's commercially available oil-sand technologies (i.e., values in Table 4.3) and based on a range of additional costs/unit of output for application of CCS (Table 4.4) to each of the two major oil-sand technologies (i.e., values in Table 4.5). We note that a number of factors could cause variation from IPCC estimates. For example, retrofitting for capture at point sources of existing oil-sand operations is a possibility. This implies costs on the high end of IPCC estimates for a number of reasons, including the existence of smaller point sources, which would not benefit from economies of scale. Additionally, switching from natural gas to bitumen or bottoms to avoid the risk associated with natural gas–price volatility would lead to a higher CO_2 intensity associated with heating and power for oil-sand operations.[38] Conversely, efficiency improvements would tend to decrease the CO_2 intensity of oil-sand operations, which would moderate this effect. Our range around the IPCC representative values is intended to capture some of this uncertainty.

---

[38] This assumes that nuclear power will not be used for heat and power applications.

Table 4.6
Expected Production Costs/Barrel of Synthetic Crude Oil in 2025

Cost	Low-Cost Assumptions (2005 $/bbl SCO)	High-Cost Assumptions (2005 $/bbl SCO)
Mining and extraction + upgrading		
Estimated 2025 production cost without CCS	27.70	33.50
CO_2 capture, transport, storage	2.90	6.40
Estimated 2025 production cost including CCS	30.60	39.90
In situ + upgrading		
Estimated 2025 production cost without CCS	30.70	35.90
CO_2 capture, transport, and storage	4.40	8.70
Estimated 2025 production cost including CCS	35.10	44.60

CHAPTER FIVE

Coal-to-Liquids Production

This chapter considers the production of CTL fuels via indirect liquefaction, which is the method for producing liquid fuels from coal that is currently receiving the most attention in the United States.[1] The analysis includes a discussion of the U.S. coal resource base, the fundamental technologies involved in CTL, the potential CO_2 emissions—based on very preliminary plant designs—and options for capturing and storing plant-site emissions of CO_2, and possible scenarios and limiting factors with respect to the development of a CTL industry in the United States.

The Coal Resource Base Relative to Coal-to-Liquids Production Needs

The United States possesses vast coal resources. Estimates of proven coal reserves are approximately 270 billion to 275 billion tons (EIA, 2006; Task Force on Strategic Unconventional Fuels, 2007). In 2005, the United States mined slightly more than 1.1 billion tons of coal, the vast majority of which (more than 1.0 billion tons) was used to produce electricity in coal-fired power stations. By 2025, the United States is expected to mine 1.5 billion short tons/year of coal, according to the EIA reference scenario in the 2007 AEO (EIA, 2007a). Bartis, Camm, and Ortiz (forthcoming) considered scenarios in which effective measures to reduce GHG emissions are adopted. They estimated an upper range of U.S. coal production (in the absence of a large CTL industry) to be 1.1 billion to 1.4 billion tons in 2030 and recognize the possibility of much lower coal production. A recent analysis of the implications of proposed legislation to reduce GHG emissions predicts significant reductions in coal production should such legislation become law (EIA, 2008a).

The development of a large CTL industry in the United States would require mining significant amounts of additional coal. A CTL facility consumes approximately 0.5 tons of coal per barrel of liquid fuel. Therefore, a plant producing 30,000 bbl/d of FT liquids requires an input of approximately 15,000 tons of coal per day. This is comparable to the amount of coal that is consumed, on average, by a 2-gigawatt conventional coal-fired power station. If, by 2025, the CTL industry produces 2 million bbl/d of coal-derived liquids, it would require approximately an additional 400 million tons of coal per year. The analysis in Bartis, Camm, and Ortiz (forthcoming) suggests that recoverable U.S. coal reserves would be sufficient to

---

[1] Much of the material in this chapter relies on Bartis, Camm, and Ortiz (forthcoming). For more detail, see Bartis, Camm, and Ortiz (forthcoming, Chapter Three) and NETL (2007c).

meet demand for electricity and the production of 3 million bbl/d of coal-derived liquids for decades past the 2025 time frame of this study.[2]

Liquid-Fuel Production via Indirect Liquefaction of Coal

A highly simplified schematic of the FT CTL process is shown in Figure 5.1. The process begins with gasifying coal, which consists of reacting coal with steam and oxygen at elevated temperatures (approximately 2,500 degrees F) and moderate pressures (400 to 500 psia) to produce a mixture of syngas and CO_2 (NETL, 2007c). As it leaves the gasifier, the syngas is dirty, because it contains CO_2 and various other gaseous molecules that derive from the impurities found in coal. These impurities would harm the performance of subsequent processing steps and are therefore removed.[3] To properly prepare the syngas, sulfur compounds are reduced to near-zero concentrations. In general, the captured sulfur would be converted to pure solid sulfur or sulfuric acid, both of which are marketable products as opposed to wastes. It is also during gas cleaning that trace mercury compounds would be completely removed.

One output of the syngas-cleaning process is a stream of CO_2. There are currently two options for the disposition of this stream: It can be emitted into the atmosphere, which would be the likely result if there were no requirement or incentive to manage GHG emissions, or it can be compressed for transport and sold for use in EOR operations (NETL, 2008). In the future, the captured CO_2 may be transported to a permanent storage location.

Figure 5.1
Process Schematic for Fischer-Tropsch Coal-to-Liquids Systems

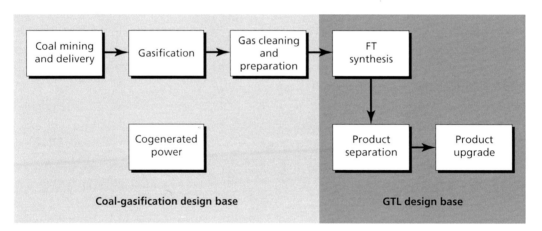

SOURCE: Bartis, Camm, and Ortiz (forthcoming).
NOTE: GTL = gas to liquids.
RAND TR580-5.1

---

[2] A recent report by the National Research Council (2007) concluded that continued expansion of U.S. coal production would benefit from updated resource estimates and R&D to improve mining productivity and reduce environmental consequences. In the time frame used for this report, these issues do not imply a threat to coal production to support unconventional-fuel development.

[3] Depending on the catalyst used in the FT synthesis reactor, preparation of the syngas may require raising the gas's H_2 content. This involves reacting a portion of the CO in the syngas with steam, which results in additional production of CO_2.

After cleaning, the syngas is directed to the FT reactors. These reactors employ catalysts of either iron or cobalt to form hydrocarbons from the syngas. The range of hydrocarbons forms a distribution of carbon lengths, from gases (such as propane and butane) to waxes (which are longer hydrocarbons). By controlling the process conditions, it is possible to shift the distribution of the hydrocarbons that the FT reactor produces (Bartis, Camm, and Ortiz, forthcoming). Additional CO_2 is typically removed at this stage of the process.

After leaving the FT reactor, the hydrocarbons are upgraded using well-established methods that are in common use in commercial refineries. The liquid fuels produced are near-zero-sulfur synthetic diesel fuel and naphtha. FT naphtha has value as a feedstock for ethylene production (NETL, 2007d). It also can be upgraded to gasoline suitable for use in automobiles. The analysis in the following section assumes that the naphtha is upgraded to reformulated gasoline.

Typical current energy efficiencies for FT CTL plants are approximately 50 percent (NETL, 2007c; SSEB, 2005). At an efficiency of 50 percent, an FT CTL plant converts one ton of bituminous or subbituminous coal (after drying) into roughly two barrels of FT liquids. Plant efficiency is a function of plant design, the type of coal fired, and plant size. Larger plants (those with production capacities above 20,000 to 30,000 bbl/d of liquid fuels) have improved scale economies and tend to be more efficient than smaller plants (those with production capacities less than 10,000 bbl/d of liquid fuels). In CTL plants that compress CO_2 to be transported and injected below ground, additional electricity would be consumed, slightly reducing overall plant efficiency.

FT CTL benefits from recent experience in the two fundamental underlying technologies: coal gasification and FT synthesis. FT CTL plants share gasification technology with IGCC electricity plants and coal-gasification facilities for the production of chemicals. A recent survey documented the construction of 13 new coal-gasification facilities between 1993 and 2004 (NETL, undated). These plants incorporate many of the same steps described already for CTL facilities, including

- continuous high-pressure gasification of coal
- handling and disposing of the ash or slag
- gas cleaning
- by-product preparation
- preparing the syngas for the FT synthesis, including possibly adjusting the ratio of H_2 to CO.

In addition to coal, other carbonaceous feedstocks may be gasified and converted to liquid fuels—in particular, biomass. This possibility has implications for the CTL plant's life-cycle GHG emissions. Recent experiments in cofeeding coal and biomass into IGCC plants have been successful (Amos, 2002; van Dongen and Kanaar, 2006).

FT synthesis is also a key component of plants that produce liquid fuels from natural gas–to-liquids (GTL) plants. Shell built a first-of-a-kind GTL facility in Malaysia in 1990 that continues to produce approximately 14,700 bbl/d (Bartis, Camm, and Ortiz, forthcoming). More recently, Sasol converted a CTL facility to accept natural gas from Mozambique. In Qatar, a large GTL facility has recently begun operating, and a second plant is under construction.

Higher oil prices have motivated further R&D to improve FT synthesis. Leaders include large firms, such as Chevron, ExxonMobil, Sasol, and Shell, and smaller firms, such as Rentech and Syntroleum. In the United States, several CTL and coal/biomass-to-liquids (CBTL) facilities have been publicly announced. Table 5.1 lists several announced CTL projects that have advanced to the front-end engineering design (FEED) phase. The FEED phase is the beginning of a facility's site-specific engineering design with a sufficient level of detail to estimate system construction and operating costs and environmental emissions and effects. Completion of the FEED phase is a critical first step in resolving cost, technical, and environmental uncertainties and leads to project financing, permitting, and the selection of a contractor to perform detailed engineering design and to construct the facility. That FEED is under way, however, does not guarantee that a plant will be constructed.

Methanol-to-Gasoline

An alternative near-term method for producing transportation fuels via the gasification of coal is the MTG approach. The MTG approach has three basic steps: Coal is gasified to produce syngas, the syngas is converted to methanol, and the methanol is converted into gasoline (Bartis, Camm, and Ortiz, forthcoming). MTG may be an attractive option for CTL in the United States because nearly all of the product slate is gasoline.[4] The liquefied petroleum gases produced at the plant may be sold or could be used at the plant to make more power or increase gasoline output. Overall plant efficiencies and CO_2-capture rates should be similar to those for FT CTL plants (Bartis, Camm, and Ortiz, forthcoming).

Current experience with components of the MTG process indicates that the technology is ready for commercial development (Bartis, Camm, and Ortiz, forthcoming). The front end of an MTG CTL facility would be very similar to that of an FT CTL facility and would take advantage of advances in coal gasification and synthesis-gas processing. Recent experience at the Eastman Chemical Company's Kingsport, Tennessee, facility demonstrates the viability of producing commercial quantities of methanol by gasifying high-sulfur bituminous coal and catalytic reaction of the syngas. A 14,500-bbl/d MTG facility using natural gas as a feedstock operated from 1985 to the mid-1990s in Montunui, New Zealand. Additionally, two MTG

Table 5.1
Proposed U.S. Coal-to-Liquids and Coal/Biomass-to-Liquids Plants

Firm	Location	Capacity (bbl/d)	Feedstock	Notes
Rentech	Natchez, Miss.	1,600 (phase 1) 28,000 (phase 2)	Coal, petroleum coke	Phase 1: demonstration; phase 2: full capacity
Baard Energy	Wellsville, Ohio	50,000	Coal, biomass	First phase would be at reduced capacity
WMPI	Frackville, Pa.	5,000	Waste anthracite	
DKRW Energy	Medicine Bow, Wyo.	15,000–20,000	Coal	MTG facility

SOURCES: Rentech (2008); Ohio River Clean Fuels (2007); DKRW Advanced Fuels (undated).
NOTE: These plants have advanced to the FEED phase.

---

[4]   Plants may also be designed to produce a small amount of electricity or LPG.

projects are under way in China. DKRW Energy has announced that it has initiated a FEED for an MTG plant to be constructed in Medicine Bow, Wyoming.

Potential Constraints on Production of Coal-to-Liquid Fuels

The price of conventional oil at which CTL-derived fuels will be competitive in 2025 depends on the experience gained in producing these fuels. In general, the more fuel that is produced, the more experience that is gained, and thus the more opportunities for cost-reducing learning that occur. This section considers some of the technical and economic challenges that may constrain the buildup of a CTL industry in the United States and potentially limit these opportunities for learning. We limit ourselves to challenges internal to the design, construction, and operation of CTL facilities. We do not consider such constraints as permitting or challenges in the expansion of coal mining and transport, among others.

CTL production based on FT synthesis uses underlying technologies that are commercially proven. However, there is a significant possibility that first-of-a-kind FT CTL plants built in the United States will experience performance shortfalls or other operational problems, especially during their initial operating years (Merrow, Phillips, and Myers, 1981). Many new U.S. CTL plants will be pioneer process plants because they will use new configurations of existing technologies drawn from the IGCC and GTL technology bases.

Challenges have surfaced in the evolving technology bases underlying FT CTL that illustrate this point. For example, the type of gasifier that is used in Sasol's major CTL facility would be considered obsolete for new CTL applications (Bartis, Camm, and Ortiz, forthcoming). There is evolving commercial experience with new gasifiers and related waste-recovery and gas-cleaning systems. Nevertheless, FT CTL facilities are very complex, and this complexity also poses risks. A case in point is the unanticipated difficulties that Sasol experienced in starting up its FT GTL plant in Qatar. In particular, Sasol experienced performance shortfalls in the FT-synthesis section of the plant during start-up operations, which extended the start-up period to more than 18 months, significantly increasing start-up costs (Sasol, 2007b). Sasol attributes the delay to challenges related to first-of-a-kind facilities (Sasol, 2007a).

FT CTL plants built in the United States will be as technically complex and challenging with similar performance risks. Recent publicly available reports on CTL assess potential development paths that could bring the United States from 0 bbl/d of CTL production to several million bbl/d of production. The Southern States Energy Board published an aggressive production-growth scenario of 1 million bbl/d of CTL by 2016, growing to 2 million bbl/d by 2019 and 5.6 million bbl/d by 2030 (SSEB, 2005). The National Coal Council (2006) proposed ramping up CTL production capacity to 2.6 million bbl/d by 2025. These estimates do not take into account cost-factor inflation or the commercial availability of CCS technology. Bartis, Camm, and Ortiz (forthcoming), alternatively, derived an estimate of the maximum production capacity for CTL, taking into account experience-based learning and the availability of commercial-scale CCS. For 2025, they estimated a potential production capacity of 1.5 million bbl/d, growing to 3 million bbl/d in 2030 (Bartis, Camm, and Ortiz, forthcoming).

In the analysis that follows, we use the Bartis, Camm, and Ortiz (forthcoming) estimate to investigate possible effects of experience-based learning on lowering production costs. We also use a low-production case for CTL-production estimates from the 2007 AEO (EIA,

2007a), with U.S. CTL-production capacity of 220,000 bbl/d, to represent a scenario in which learning-based cost reduction is negligible.

Other key uncertain factors that have a significant bearing on the potential capacity of CTL are the trajectory of world oil prices, actual production costs of CTL, and the availability of commercial-scale CCS (Bartis, Camm, and Ortiz, forthcoming). Environmental concerns related to significant increases in coal mining would also need to be considered. Current designs for CTL facilities consume significant amounts of water, which may pose some technical and resource challenges, especially in the western United States. In principle, these problems can be overcome.

Carbon-Dioxide Production and Capture for Coal-to-Liquids

Baseline Carbon-Dioxide Emissions from Coal-to-Liquids Production

Many view the potentially high emissions of GHGs at the CTL plant site to be a barrier to the development of a CTL industry in the United States. The life-cycle GHG emissions of CTL production are approximately twice those of conventional fuels on a per-unit basis.[5] In the absence of CCS, these emissions include CO_2 produced during gasification and FT synthesis and fuel-combustion and other GHGs released, such as coal-mine methane. The CTL facilities in the FEED stage of development listed in Table 5.1 all propose to capture and store plant-site CO_2 emissions. In general, the amount of CO_2 released at CTL plants would be approximately 0.8 tons/bbl of product (Bartis, Camm, and Ortiz, forthcoming). In our hypothetical CTL industry producing 1.7 million barrels/day in 2025, these plant-site emissions would be approximately 1.4 million tons of CO_2/day.

Mixing Biomass and Coal to Reduce Coal-to-Liquids Carbon-Dioxide Emissions

There are two methods for reducing the CO_2 emissions of CTL production: constructing an FT facility to accept a mixture of coal and biomass[6] as a feedstock and employing CCS systems. Applied separately, each of these two approaches can bring life-cycle CO_2 emissions to a level comparable to those of conventional petroleum. An integrated plant designed to accept both coal and biomass can lower life-cycle emissions of CO_2. The use of biomass reduces the plant's net CO_2 emissions, taking into account the CO_2 removed from the atmosphere in growing the biomass.[7] This dual-fired approach can also offer improved plant operating characteristics over

---

[5]   This estimate depends on several assumptions: The crude oil is light (35 degrees or more on the API gravity scale); refinery efficiencies are 87 percent for low-sulfur diesel and 85.5 percent for reformulated gasoline; the CTL plants are those of SSEB (2005, cases 3 and 8), which are relatively inefficient; CH_4 emissions from coal mining are consistent with estimates from emissions reported in EIA (2007e). These assumptions apply to a case in which CTL production without CCS is compared with imported oil. It is also possible to compare CTL production with heavier crudes, which have greater life-cycle emissions of CO_2 and would result in a lower range of values.

[6]   As defined by the U.S. Department of Energy (undated), "the term 'biomass' means any plant-derived organic matter available on a renewable basis, including dedicated energy crops and trees, agricultural food and feed crops, agricultural crop wastes and residues, wood wastes and residues, aquatic plants, animal wastes, municipal wastes, and other waste materials."

[7]   GHG emissions of sustainably regenerated biomass are associated only with cultivation, harvesting, and delivery of the biomass to the plant. If an FT plant were to operate on a biomass feed alone, without employing systems for CCS, the life-

relying just on biomass. The inclusion of coal offers a consistent feedstock supply that can balance fluctuations in the availability of biomass and enable economies of scale not available if the plant were supplied by biomass alone.

Several recent studies have investigated the characteristics of CBTL plants. Among others, Williams, Larson, and Jin (2006) investigated the cofiring of coal and biomass using feedstocks that require few or no agricultural inputs. These feedstocks are mixtures of perennial grasses, legumes, and trees and have been shown to reconstitute soil carbon content as they produce harvestable biomass (Tilman, Hill, and Lehman, 2006). The additions to soil carbon would be another benefit of using these crops. The National Energy Technology Laboratory recently completed an engineering analysis of the use of corn stover, farmed poplar trees, and switchgrass with coal to produce liquid fuels (NETL, 2007d).

Several recent analyses have investigated the potential biomass resource that exists in the United States. A study from Oak Ridge National Laboratory (Perlack et al., 2005) estimated that approximately 330 million dry tons per year of forest residue and thinnings, urban wood waste, and agricultural residue are available annually, assuming no changes in land use or increases in agricultural yields, and up to 1.4 billion tons annually with more-intensive collection efforts, increased yields, and land-use changes. This estimate is based solely on land availability and does not consider the cost of collecting and delivering biomass to sites where it would be used to produce electric power, useful chemicals, or liquid fuels. Assuming modest changes in agricultural yields and cultivation of energy crops on idle land, the National Renewable Energy Laboratory (Milbrandt, 2005) performed a similar analysis and estimated that 360 million dry tons per year of such biomass resources exist. Assuming that 700 million tons per year (half of the high Oak Ridge National Laboratory estimate) can be economically collected and delivered to liquid fuel–production facilities, biomass resources (without coal blended in) would be able, in principle, to support a fuel-production level of roughly 3 million bbl/d (Bartis, Camm, and Ortiz, forthcoming).[8]

Recent results, however, call into question the magnitude of potential CO_2 reductions from biomass-derived fuels. When production of a biomass feedstock displaces active farmland, it induces land-use change to make up for the lost food production. When this land, such as tropical rain forest or grassland, contains a significant amount of stored carbon, that carbon is released when the land is prepared and tilled for planting. Fargione et al. (2008) and Searchinger et al. (2008) addressed this issue. As part of an analysis supporting the California renewable-fuel standard, Farrell and O'Hare (2008) have also considered the effect of land-use changes on CO_2 emissions.[9] All three of these reports concluded that the CO_2 released when lands are switched from other uses to producing traditional biofuel crops can dominate the life-cycle emissions of those fuels for a considerable period (e.g., decades) after the land conversion. It is also possible to run this process in reverse, growing crops specifically to store carbon in the soil (Tilman, Hill, and Lehman, 2006). Growing biomass on degraded land, however, can result in lower yields. Additional uncertainties exist regarding the appropriate methods

---

cycle emissions would be only 10 to 15 percent of those associated with conventional petroleum-based fuels.

[8]   However, this would preclude the use of any of this nonagricultural biomass for lower-CO_2 electricity production. Toman, Griffin, and Lempert (2008) analyzed this trade-off.

[9]   The California Air Resources Board (2008) maintains a Web site that contains all relevant documentation and analyses pertaining to the state's low-carbon–fuel standard.

for accounting for emissions of N_2O resulting from fertilization (Crutzen et al., 2007) and the effects of alternative farming practices (Baker et al., 2007).

Carbon Capture for Coal-to-Liquids

CCS is the second option available to reduce the CO_2 emissions from CTL production. If the plant-site emissions of a CTL facility can be successfully captured and stored, the life-cycle emissions from producing and using the FT fuels will be comparable to those of conventional fuels. CO_2 capture for CTL facilities is covered in detail in Bartis, Camm, and Ortiz (forthcoming).

The development of CTL capacity in the United States would result in a potentially large supply of CO_2 for permanent storage. As discussed, 85 to 90 percent of plant-site CO_2 emissions are removed from process streams while preparing the syngas for FT synthesis. To prepare the CO_2 for transport and storage requires only that the CO_2 stream be dehydrated and compressed. These emissions may be captured at a low cost, at least relative to the costs of employing CCS in new coal-fired power plants (Bartis, Camm, and Ortiz, forthcoming). The remaining sources of CO_2 emissions at a CTL plant are due to the combustion of fuel gases, which are used to produce electric power to support plant utilities. These fuel gases are produced as part of the distribution of hydrocarbons in FT synthesis and during product-upgrading steps that occur within the plant.[10] For FT CTL plants, such as the one analyzed in the following section, emissions from fuel-gas combustion can represent between 10 and 20 percent of the plant's site CO_2 emissions (Bartis, Camm, and Ortiz, forthcoming). For example, the type of CTL plant examined in this section captures 0.74 tons of CO_2 per barrel of liquid fuels produced and emits 0.082 tons of CO_2 per barrel of liquid fuels produced. Capturing emissions from fuel-gas combustion would be relatively expensive (Bartis, Camm, and Ortiz, forthcoming) and is not considered in the following analysis.

Potential Future Unit Production Costs for Coal-to-Liquids

In this section, we estimate the production costs of CTL in 2025. Our approach is to start with production-cost estimates for first-of-a-kind CTL facilities that could occur within the next 10 years, based on the analysis in Bartis, Camm, and Ortiz (forthcoming). We then adjust those production costs for uncertainties in future capital costs and experience-based learning out to 2025. We also calculate a CO_2 balance for first-of-a-kind plants to determine the emissions of CO_2 per unit of FT product. This allows us to evaluate the cost of the FT fuel including CO_2-emission costs. We consider ranges of cost estimates that are plausible given available information regarding the technology status of and key inputs to CTL facilities. Given the considerable uncertainties in predicting production costs, the ranges should not be interpreted as upper and lower bounds.

With respect to production-cost uncertainties, prior RAND research (Merrow, 1989; Hess and Myers, 1988; Merrow, Phillips, and Myers, 1981; Myers and Arguden, 1984; Myers, Shangraw, et al., 1986) studied the phenomenon of cost growth and performance shortfalls in pioneer process plants. This research concluded, "Most of the variation found in cost-estimation

---

[10] The fuel gas consists of a mixture of nitrogen, H_2, CO, and light hydrocarbons, such as CH_4, propane, and butane (NETL, 2007c).

error can be explained by (1) the extent to which the plant's technology departs from that of prior plants, (2) the degree of definition of the project's site and related characteristics, and (3) the complexity of the plant" (Merrow, Phillips, and Myers, 1981).[11] When applying these general results to CTL, we observe the following:

- The technology derives from an established base in IGCC and GTL.
- Current plant designs incorporate little definition with respect to the site.

The phenomenon of cost growth is most prevalent in pioneer—i.e., first-of-a-kind—process plants. With a more mature CTL industry—albeit one in which there are still important opportunities for technological improvement—the key uncertainties in the construction of a new plant would be site-specific issues, such as the availability of process water, feedstock and product transportation, and environmental compliance. We deal with this possibility by postulating a larger contingency in the estimated capital cost, as discussed later. We also allow for potential declines in the future cost of CTL product due to experience-based learning.

We start with a characterization of a baseline first-of-a-kind facility that has a nominal capacity of 30,000 bbl/d of FT liquids; approximately two-thirds is FT diesel fuel and one-third is FT naphtha (see also SSEB, 2005, case 3). This is the plant considered in Bartis, Camm, and Ortiz (forthcoming). Those authors converted the production of naphtha to a price-equivalent production of FT diesel—*diesel value equivalent* (DVE)—to facilitate analysis and interpretation. Electricity is a required plant utility, and this plant also produces 204 MW for export. We assume that this electricity is exported as base-load power for a price of $0.05/kilowatt-hour (kWh).

Table 5.2 summarizes the key aspects of the unit-cost analysis for this first-of-a-kind plant for both a base case and a high-cost case. The base and high-cost cases differ in the estimates of capital costs and fixed (nonfuel) operating costs. Capital costs include two basic categories: plant costs and start-up costs. Plant costs dominate in the calculations and include all site preparation; design and construction costs for the entire plant, including on-site product upgrading required to produce a finished diesel fuel and a naphtha product suitable for pipeline transport to a refinery; auxiliary systems required for pollution control and coal-ash disposal; and infrastructure required to obtain access to electric power, water, and coal and to transport the products that the plant creates. For CTL plants built in the West, plant costs could include substantial additional investments in infrastructure, such as roads, housing of construction workers, and water access. See Bartis, Camm, and Ortiz (forthcoming, Appendix A) for details of the financial analysis of the facility.

For both base and high-cost cases, we derive production costs with and without CCS capability. CCS cost estimates for CTL facilities also are drawn from Bartis, Camm, and Ortiz (forthcoming). According to Gray and White (2007), since CO_2 capture is an integrated part of the plant processes, if the plant were to emit the captured CO_2, there would be an additional 53 MW of electricity available for export. A plant emitting CO_2 would also have no need for equipment capable of compressing the captured CO_2 for transportation and storage. The capital costs for the no-CCS case presented in Table 5.2 have been adjusted to exclude

---

[11] Merrow, Phillips, and Myers (1981) defined *complexity* as a count of all major process blocks.

Table 5.2
Technical and Economic Parameters for a First-of-a-Kind Coal-to-Liquids Plant

Parameter	With CCS Capability		Without CCS Capability	
	Base Case	High-Cost Case	Base Case	High-Cost Case
Plant coal feed (tons/d of bituminous coal)	17,987	Same	Same	Same
Plant production capacity (bbl/d DVE)	32,502	Same	Same	Same
Plant capacity factor	0.9	Same	Same	Same
Economic life of facility (years)	30	Same	Same	Same
Plant capital costs, base values ($ billions)	3.31	4.05	3.12	3.82
Annual fixed operating costs ($ millions)	132	165	132	165
Variable operating costs ($/bbl DVE)	2.69	Same	Same	Same
Cost of coal ($/ton)	30	Same	Same	Same
Electricity for export (MW)	204	204	257	257
Sale price of exported electricity ($/kWh)	0.05	0.05	0.05	0.05

SOURCES: Bartis, Camm, and Ortiz (forthcoming); SSEB (2005, case 3).
NOTE: Figures are in 2007 dollars.

this equipment.[12] The figures with CCS capability do not include the costs of transport and storage of the captured and compressed CO_2. These costs are described later.

Table 5.3 presents production-cost components per gallon (gal.) of FT diesel produced. Total production costs in 2007 dollars range from $1.56/gal. in the base case without CCS to $2.00/gal. in the high-price case with CCS. The $0.11 to $0.13 per-gal. difference in production cost between the with- and without-CCS cases are consistent with the estimates in Bartis, Camm, and Ortiz (forthcoming). To compare these estimates with those of oil sands presented in the previous chapter and with 2007 projections from EIA, we convert these figures to 2005 dollars using the general gross domestic product (GDP) deflator, which is 1.064 (comparing first-quarter price indices) (BEA, 2008).

Carbon Dioxide–Management Cost for CTL

To determine the cost of CO_2 capture, we draw on the analysis of Bartis, Camm, and Ortiz (forthcoming, Appendix B). We assume that 90 percent of the CO_2 that would otherwise be emitted into the atmosphere is captured and compressed; this is the upper end of the range of potential capture rates discussed earlier. Additionally, Bartis, Camm, and Ortiz (forthcoming) assumed that the FT naphtha produced at the plant would be upgraded to reformulated gasoline. The hypothetical facility analyzed in SSEB (2005, case 3) captures 7.7 million tons of CO_2 annually.

Emissions of CO_2 for SSEB (2005, case 3) are compared to emissions of CO_2 from conventional fuels. Table 5.4 (from Bartis, Camm, and Ortiz, forthcoming) lists life-cycle CO_2

---

[12] Specifically, we exclude from the capital costs the CO_2-capture component of capital costs from SSEB (2005, Appendix D). This block includes some plant functions in addition to compressing the CO_2, which induces a small amount of error in our estimate of production costs.

emissions from conventional fuels and from the FT facility considered in this analysis. The emissions of

Table 5.3
Estimated Component Costs per Unit of Production from First-of-a-Kind Coal-to-Liquids Plants

Component of Product Cost	With CCS Capability		Without CCS Capability	
	Base Case	High-Cost Case	Base Case	High-Cost Case
Capital charge ($/gal. diesel)	1.09	1.35	1.03	1.27
Fixed operating cost ($/gal. diesel)	0.30	0.38	0.30	0.38
Variable operating cost ($/gal. diesel)	0.06	0.06	0.06	0.06
Coal ($/gal. diesel)	0.39	0.39	0.39	0.39
Electricity sales ($/gal. diesel)	−0.18	−0.18	−0.23	−0.23
Total cost ($/gal. diesel) (2007 $)	1.67	2.00	1.56	1.87
Total cost ($/gal. diesel) (2005 $)	1.57	1.88	1.47	1.76

SOURCE: Bartis, Camm, and Ortiz (forthcoming).

NOTE: Figures are in 2007 dollars unless otherwise noted. Figures may not sum due to rounding. Costs of CO_2 transport and storage are not included in figures for CTL production with CCS capability.

Table 5.4
Comparison: Life-Cycle Greenhouse-Gas Emissions of Conventional Fuels and Synthetic Fuels from a Hypothetical Fischer-Tropsch Facility

Life-Cycle Step	Low-Sulfur Diesel (lb CO_2e/ mmBtu)	Reformulated Gasoline (lb CO_2e/ mmBtu)	Weighted Average Conventional (lb CO_2e/ mmBtu)	CTL Without CCS (lb CO_2e/ mmBtu)	CTL with CCS (lb CO_2e/ mmBtu)	
					Electricity Credit	No Electricity Credit
Extraction and mining	12.5	12.7	12.5	37.8	37.8	37.8
Feedstock transportation	2.0	2.0	2.0	0.0	0.0	0.0
Refining and production	26.2	29.6	27.0	309.7	31.0	31.0
Credit for exported electricity	0.0	0.0	0.0	−55.5	−57.5	0.0
Product transportation	1.1	1.0	1.0	1.0	1.0	1.0
Product combustion	163.0	157.7	161.6	156.3	156.3	156.3
Total	204.7	203.0	204.2	449.4	168.7	226.2
CO_2e emissions (tons/bbl product)	0.60	0.52	0.58	1.2	0.45	0.60
CO_2 stored (tons/bbl product)	0.00	0.00	0.00	0.00	0.74	0.74

SOURCE: Bartis, Camm, and Ortiz (forthcoming).

NOTE: Emissions from refining for CTL production are based on SSEB (2005, case 3), which is a nominal 30,000-bbl/d facility producing liquid fuels and electricity from bituminous coal. mmBtu = millions of British thermal units.

CO_2 for oil recovery, coal mining, transportation of resources, and final products, are drawn from GREET model version 1.8a (Argonne National Laboratory, 1999). Several assumptions are made to facilitate comparison. For cases without CCS, exported electricity is credited as if it were produced at an IGCC facility with a heat rate of 7,900 Btu/kWh. For cases with CCS, exported electricity is credited at the U.S. average heat rate, in 2005, of 10,400 Btu/kWh. For comparison, a case is provided in which no credit is given. Emissions from fuel transportation are assumed to be the same as in the conventional case. A weighted average of conventional fuels, based on 74-percent diesel and 26-percent reformulated gasoline by energy, is tabulated to facilitate comparison with the output of the FT facility.

Two additional components of CO_2 management are transportation from the plant and storage in a geologic reservoir. We adopt a range of potential costs for transportation and storage from IPCC (2005, Table 8.1).[13]

Potential Cost Declines from Learning

To assess possible reductions in production costs due to learning, we consider the possibility that base-case capital and fixed-operating-cost components for first-of-a-kind plants in Table 5.3 could decline 20 percent by 2025. We can justify this decline by assuming that these costs decline by 10 percent for each doubling of cumulative output, based on a potential trajectory for cumulative production of CTL through 2025 in Bartis, Camm, and Ortiz (forthcoming). This is the same construct used for oil sands in Chapter Four. We use the accelerated CCS demonstration case from Bartis, Camm, and Ortiz (forthcoming, Table 3.1), adjusting the total capacity downward to reflect our assumed 90-percent capacity factor.

In contrast, we can consider what could happen if the sector developed more slowly and learning was correspondingly slower. In that case, we would have cost figures much more akin to those shown in Table 5.3, which are based on no learning-based improvement. To construct our high-cost case for 2025, accordingly, we use the high-cost case in Table 5.3.

Table 5.5 provides the 2025 unit costs used in the next chapter to compare CTL with conventional petroleum fuels under different assumptions. The production costs with learning are derived by applying the assumed 20-percent cost savings to capital and fixed operating costs in Table 5.3. Table 5.5 also includes the additional costs for CO_2 transport and storage in the CCS cases; the costs of storage and monitoring are the same as in Table 4.4 in Chapter Four.

Table 5.5
Alternative Coal-to-Liquids Unit Production Costs for 2025

Cost (2005 $/gal. diesel)	With CCS		Without CCS	
	Base Case with Learning	High-Cost Case Without Learning	Base Case with Learning	High-Cost Case Without Learning
Estimated 2025 production cost	1.31	1.88	1.22	1.76
CO_2 transportation and storage	0.07	0.35	0.00	0.00
Estimated 2025 production cost including CCS	1.38	2.23	1.22	1.76

SOURCE: Production cost is based on Table 5.3. CO_2 transportation and storage costs are from Chapter Three.

---

[13] We have used the range of costs for a 250-km pipeline at 7.3 Mtonne CO_2/yr (8.0 Mton CO_2/yr), which is approximately 2002 US$1.70–3.00/tonne CO_2, and converted these to 2005 US$/ton CO_2 per the method in Chapter Four. This is based on a 33,000 bbl/d facility and storage requirements at 0.74 ton CO_2/bbl FT liquids.

CHAPTER SIX

Competitiveness of Unit Production Costs for Synthetic Crude Oil and Coal-to-Liquids

In this chapter, we describe how the unit costs of both SCO and CTL would compare to the prices of products derived from conventional crude under different assumptions about the production costs of these unconventional fossil-based fuels and the cost of CO_2 emissions. Since products derived from oil sands and CTL are being compared to conventional crude-oil products, we must also account for the uncertainty that surrounds future world oil prices and apply assumed costs of CO_2 emissions to conventional fuels.

In this analysis, we look at a range of potential future oil prices in 2025 taken from the 2007 AEO (EIA, 2007a). We focus particularly on the reference-case and high-oil-price scenarios, with the latter being of interest given concerns about the potential for persistently high oil prices. However, we also touch on the implications for SCO and CTL of lower oil prices, in the range of EIA's low-oil-price scenario.

The ranges of unit costs for SCO and CTL reflect the analyses in Chapters Three through Five. We consider a range of possible costs of CO_2 emissions. We also consider different potential future costs of CCS as applied to producing alternative fuels. All fuels, including those derived from conventional petroleum, are subject to a cost applied to the emissions from final fuel consumption as well as emissions from production (with or without CCS). Thus, we consistently apply the CO_2-emission cost on a full life-cycle basis to conventional petroleum as well as to the unconventional fossil-based fuels. The emission costs for SCO and fuels from CTL reflect the relative life-cycle CO_2 intensity of these fuels.

The relative CO_2-intensity assumptions underlying the different scenarios that we consider for the unit-cost comparisons are summarized in Table 6.1. For oil sands, we use the ratio of life-cycle emissions of CO_2 from SCO and conventional crude oil. For CTL, the ratio we use is that of life-cycle emissions of CO_2 per unit of product and a mixture of conventional low-sulfur diesel and reformulated gasoline. To compare SCO directly with CTL would require including refinery losses involved in converting SCO to a comparable slate of diesel fuel and reformulated gasoline, which we do not undertake here.

Estimates of energy-security costs associated with conventional petroleum cover a wide range, representing many assumptions. A recent study by Leiby (2007), using a wide range of assumptions for oil-market conditions and macroeconomic vulnerabilities, concluded that the added social cost of one more barrel of oil imports and consumption was in a range of about $0.20 to more than $0.33/gal. These figures include the cost of oil exporters maintaining prices above competitive levels and expected macroeconomic costs from future oil-price spikes. When one deducts the macroeconomic vulnerability components, following the argument that these figures do not apply to a relatively small fuel substitution, as noted in Chapter Two, the

Table 6.1
Comparison: Life-Cycle Greenhouse-Gas Emissions for Unconventional Fossil-Based Products Relative to Conventional Low-Sulfur, Light Crude Oil

Technology	Ratio of GHG-Emission Intensity	
	Low	High
Oil-sand (SCO to crude)		
Mining, no CCS	1.13	1.22
Mining with CCS	0.98	0.99
SAGD, no CCS	1.19	1.26
SAGD with CCS	0.98	0.99
CTL (CTL fuel to conventional liquid fuel)		
CTL, no CCS	2.2	NA
CTL with CCS	0.82	1.1

NOTE: Ranges for oil sands are based on observed variability of oil-sand production facilities, as reported in Chapter Four. Ranges for CTL are based on the per-unit-of-energy emissions of CO_2 reported in Chapter Five. The conventional-petroleum baseline is a light crude oil. The conventional-fuel baseline is a mixture of low-sulfur diesel fuel and reformulated gasoline in the same proportions as produced by the CTL facility.

range declines to about \$0.10–\$0.20/gal., or roughly \$4.20–\$8.40/bbl.[1] The study by Bartis, Camm, and Ortiz (forthcoming) of CTL calculates a range of benefits from \$3.70–\$18.30/bbl of oil equivalent of alternative fuels under 2007 AEO reference-case assumptions and \$5.50–\$29.30/bbl for the 2007 AEO high-oil-price case. These figures assume 3 million bbl/d of U.S. CTL production. The upper ends of these ranges assume a less cohesive OPEC that has difficulty maintaining prices in the face of increased competition and a low elasticity of oil demand, so that increased competition can drive prices down more.

Oil Sands

For oil sands, we carried out analyses for both of the technologies discussed in Chapter Four—surface mining and in situ production via SAGD. As discussed previously, the opportunities to apply CCS in the production of oil sands occur at points related to extracting and upgrading the bitumen to SCO (e.g., centralized heaters, power generation). Four sets of technology and cost assumptions are considered for each of the two production technologies, relative to a base-case set of anchoring parameters reflecting the analysis in Chapter Four: high and low cases for production prices and high and low cases for CO_2 intensities and CCS costs.[2] These

---

[1] Even lower numbers would apply if energy demand and non-OPEC supplies were more responsive than Leiby assumed, so the market power to be countervailed would be lower.

[2] We also could have considered ranges of future natural-gas prices in this construction of assumption sets. As noted in Chapter Four, however, the economic sensitivity of SCO cost to natural-gas price is less than the capital-cost sensitivity. To help keep the comparisons that follow more manageable and easily interpreted, we hold the natural-gas price at the AEO 2007 reference-case level across all scenarios.

four assumption sets, summarized in Table 6.2, bound the range of technological possibilities that we consider.

Cost Comparison for Synthetic Crude Oil Produced by Integrated Mining and Upgrading

Figure 6.1 displays the results of our unit-cost comparisons for SCO produced by integrated mining and upgrading. The four colored upward-sloping lines identified in the legend represent the full unit costs of SCO with integrated mining and upgrading, corresponding to the four cases defined in Table 6.2. Each line represents the full unit cost for different values of the parametric CO_2-emission cost. We also show two upward-sloping lines in the figure to represent the price of conventional crude under reference-case and high-price-case assumptions, including the corresponding CO_2-emission cost. As noted previously, the emission cost is applied to the life-cycle emissions of CO_2.

The orange region in the figure represents the range of production costs and the range of CO_2 intensities of the technique for oil-sand production in the absence of CCS. The blue region represents the ranges of production costs and CO_2 intensities, along with the range of different costs accompanying the use of CCS. The vertical distance between the two orange lines indicates the range of possible SCO costs for different assumed capital and operating costs and CO_2 intensities without CCS, for any specified CO_2-emission cost. This distance is relatively small for most of the CO_2-emission costs shown in the figure, as one would expect of a relatively mature technology. A similar interpretation can be applied to the vertical distances between the blue lines. This distance is larger than the vertical distance without CCS, since the addition of a range of CCS costs amplifies the range of potential unit-production costs for SCO. It is useful to keep in mind, however, that our upper-bound assumption for CCS is fairly pessimistic (high cost).

Table 6.2
Oil-Sand Comparison Cases

Scenario	Description and Assumptions	Unit Cost (2005 $/bbl SCO)	
		Method	Cost
Current estimated costs	Natural gas = 2005 US$6.05/mmBtu (EIA reference value for electric power in 2025 [EIA, 2007a]) Capital and nongas operating costs = NEB 2006 assumptions, as represented in Lacombe and Parsons (2007)	Integrated mining	33.50
		In situ	35.90
Low cost, low CO_2 intensity, no CCS	−20% estimated current capital costs Low estimates of CO_2 intensity	Integrated mining	27.70
		In situ	30.70
High cost, high CO_2 intensity, no CCS	Estimated current capital costs High estimates of CO_2 intensity	Integrated mining	33.50
		In situ	35.90
Low cost with low-cost CCS and low CO_2 intensity	−20% estimated current capital costs Low estimates of CO_2 intensity Low cost estimates for CO_2 capture	Integrated mining	30.60
		In situ	35.10
High cost with high-cost CCS and high CO_2 intensity	Estimated current capital costs High estimates of CO_2 intensity High cost estimates for CO_2 capture	Integrated mining	39.90
		In situ	44.60

SOURCE: Table 4.6 in Chapter 4

Figure 6.1
**Estimated Unit Production Costs of Synthetic Crude Oil from Integrated Mining and
Upgrading of Oil Sands, with and Without Carbon Capture and Storage, and of
Conventional Crude Oil in 2025, Versus Different Costs of Carbon-Dioxide Emissions**

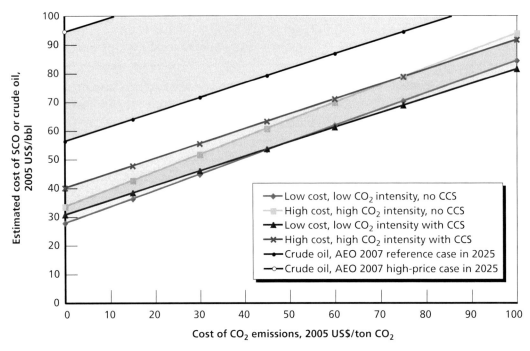

RAND *TR580-6.1*

Because oil sands are approximately 20 percent more CO_2-intensive than conventional
crude oil without CCS, as the cost of CO_2 emissions rises, the unit cost of SCO including
the CO_2-emission cost rises more quickly (along the orange lines) than the unit price of crude
including its CO_2-emission cost. The slope of the upper-bound line is greater than the slope of
the lower-bound line because the upper line has a greater CO_2 intensity.

With CCS, the full unit cost of SCO from integrated mining and upgrading is less sensitive to the cost of CO_2 emissions as a result of capturing most of the point-source emissions.
This is indicated by the flatter slopes of the lines bounding the blue region compared to those
bounding the orange region. Again, the upper blue line has a steeper slope than the lower blue
line to reflect the greater CO_2 intensity in the former case.

Figure 6.1 indicates that the technology of oil-sand mining with or without CCS appears
to be very cost-competitive with conventional crude for a wide range of CO_2-emission costs.[3]
In particular, it is very competitive given the 2007 AEO reference-case price of oil at $56/bbl
in 2025. This suggests that SCO production will be cost-competitive unless longer-term oil
prices fall back to levels well below those experienced over the past few years.[4]

---

[3]   In fact, even without CCS and under the high-capital-cost scenario (i.e., the upper line of the orange region), oil-sand
production of SCO would be profitable up to emission costs of $240/ton CO_2 (not shown in the figure). For other combina-
tions of oil prices and technology costs, the corresponding threshold for CO_2-emission costs is well into the thousands of
dollars/ton of CO_2.

[4]   Even for an oil price as low as $35/bbl (the 2007 AEO low-price-case value for 2025), SCO integrated mining without
CCS is cost-competitive if the CO_2-emission cost is relatively low.

We also can compare the boundaries of the orange and blue areas in Figure 6.1 to draw implications for the choice of adding CCS to integrated mining. In the case of low SCO-production cost, low CCS cost, and low CO_2 intensity of SCO production, compliance with a regulatory program resulting in paying a CO_2-emission cost is more cost-effective for the producer than installing CCS if the CO_2-emission cost is less than about \$50/ton. The corresponding break-even CO_2-emission cost with high costs and high CO_2 intensity of SCO production is about \$75/ton.

Now consider the decision of a developer of a future new oil-sand site regarding the installation of CCS with integrated mining. When the overall site investment must be planned and initial work undertaken, the developer will still face uncertainty about SCO-production cost and CCS costs, as well as the future CO_2-emission cost. Committing to installation of CCS will lead to a higher cost of SCO if the CO_2-emission cost is low to moderately large, especially in the high-cost case. Including CCS will also make the anticipated unit cost more uncertain. Figure 6.1 suggests that, even with high confidence that the cost of CCS will be low, this investment is not attractive for the producer unless there is also a strong expectation of high CO_2-emission costs. However, installation of CCS would have only a limited effect on the cost-competitiveness of SCO relative to crude, given our cost assumptions and the AEO 2007 prices.

Cost Comparison for Synthetic Crude Oil Produced by Steam-Assisted Gravity Drainage and Upgrading

Figure 6.2 presents the analysis for SAGD with upgrading, which is likely to be the source of incremental SCO supply with significant expansion of total output. This figure is read in the same way as Figure 6.1. SAGD with upgrading has slightly greater production costs than integrated mining and upgrading. Nevertheless, SCO produced with this technology appears to be cost-competitive with conventional oil over a wide range of CO_2-emission costs and crude-oil prices.[5] The break-even CO_2-emission cost with and without CCS are about the same as in Figure 6.1, and a commitment to install CCS has the same effect of raising production costs and increasing their uncertainty for low to moderately high CO_2-emission costs. Therefore, the economic benefit of adding CCS investment to SAGD is unclear.

Coal to Liquids

As with oil sands, we develop four sets of assumptions that bound the range of possibilities we consider. These are shown in Table 6.3. We consider base and high-cost cases for the production technology, including different assumptions about cost reduction through learning. Each cost option is considered with and without CCS. The two assumption sets with CCS also incorporate high or low CCS costs and intensities. We use the CO_2-emission intensity derived in Bartis, Camm, and Ortiz (forthcoming, Appendix B) and reported in Table 5.4 in Chapter Five of this report as the basis for comparison. We weight the CTL product outputs on an energy basis to develop a composite measure of CTL output. In particular, the output of liquid fuels from our base CTL plant, on an energy basis, is assumed to be 74 percent FT diesel and

---

[5] Without CCS, the full unit cost of SCO is lower than the full unit cost of conventional oil up to a \$180/ton CO_2 emission cost, not shown in the figure.

Figure 6.2
Estimated Unit Costs of Synthetic Crude Oil from Steam-Assisted Gravity Drainage with Upgrading of Oil Sands, with and Without Carbon Capture and Storage, and of Conventional Crude Oil in 2025, Versus Different Costs of Carbon-Dioxide Emissions

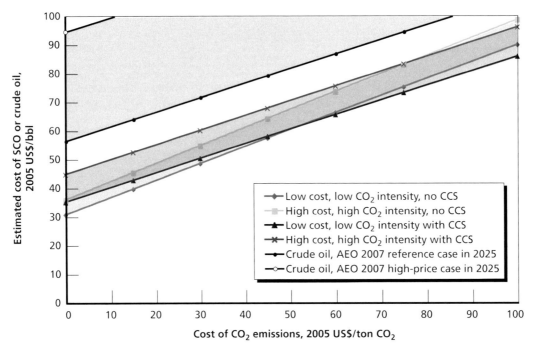

RAND *TR580-6.2*

Table 6.3
Coal-to-Liquids Comparison Cases

Scenario	Description and Assumptions	Unit Cost (2005 $/gal. diesel)
Base-case production cost; learning; low CO_2 intensity; no CCS	Current costs based on SSEB (2005, case 3); 10% cost reduction for each doubling of production; CO_2 credit for exported electricity production; CO_2 vented to atmosphere	1.22
High cost; no learning; high CO_2 intensity; no CCS	125% of base capital and operating costs; no cost reduction; no CO_2 credit for exported electricity production; CO_2 vented to atmosphere	1.76
Base-case production cost; learning; low CO_2 intensity; low-cost CCS	Current costs based on SSEB (2005, case 3); 10% cost reduction for each doubling of production; CO_2 credit for exported electricity production; 90% of plant-site CO_2 emissions captured and stored at low-cost estimate from IPCC (2005)	1.38
High production cost, no learning, high CO_2 intensity, high-cost CCS	125% of base capital and operating costs; no cost reduction; no CO_2 credit for exported electricity production; 90% of plant-site CO_2 emissions captured and stored at high-cost estimate from IPCC (2005)	2.23

SOURCE: Table 5.5 in Chapter Five.

Figure 6.3
Estimated Unit Production Costs of Fischer-Tropsch Diesel from Coal, with and Without Carbon Capture and Storage, and of Diesel in 2025, Versus Different Costs of Carbon-Dioxide Emissions

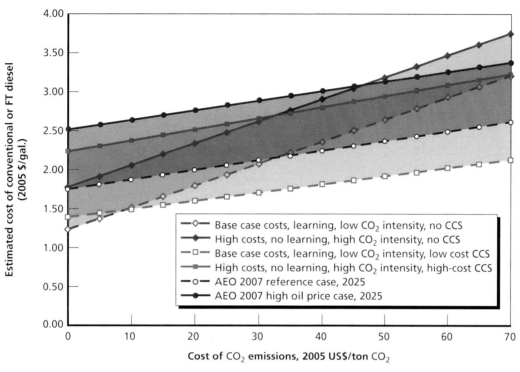

RAND *TR580-6.3*

26 percent reformulated gasoline. We compare the life-cycle CO_2 emissions of the same mix of conventionally produced diesel and reformulated gasoline.

The results for our CTL analysis appear in Figure 6.3. The red and green lines show the full unit costs of CTL products with and without CCS for the different cost and CO_2-intensity assumptions, given different CO_2-emission costs. The 2007 AEO reference and high-oil-price cases for conventional fuels (here represented by the corresponding costs for diesel fuel) are also shown. All lines slope upward with an increase in the CO_2-emission cost. The lines for CTL without CCS slope upward more steeply than the lines for CTL with CCS and more steeply than the lines for conventional fuels (reflecting the greater CO_2 intensity of CTL without CCS). Because of the high efficiency of CO_2 capture in CTL production, the green lines (CTL with CCS) are almost parallel to the lines for conventional products (for each of these lines, the upward slope reflects mainly end-use emissions). Due to the additional cost uncertainties for transporting and storing CO_2, the range of costs with CCS is larger than the case in which the CO_2 is emitted and the CO_2-emission cost is paid.

With a CO_2-emission cost of zero and without CCS, CTL cost ranges from \$1.22 to \$1.76/gal. This can be compared to the 2007 AEO reference-case price in 2025 for conventional diesel, which is \$1.74/gal. CTL thus appears to be relatively cost-competitive with conventional fuel under the assumptions indicated, though the degree of competitiveness is less than for oil sands and more sensitive to the possibility of high CTL-production costs and long-term oil prices below the EIA 2007 reference-case level. With CCS, the comparable range for

CTL is $1.38 to $2.23/gal. This range overlaps the EIA reference-case price for conventional diesel, but it is somewhat below the high-oil-price case diesel price of $2.51/gal. in 2025. Thus, while CTL is cost-competitive with conventional fuel at prices at or above the EIA high-oil-price case, it may not be cost-competitive for oil prices at or even somewhat above the reference case if CTL and CCS costs are high.

At a low cost of CO_2 emissions, there is a cost advantage to emitting the CO_2 to the atmosphere rather than compressing, transporting, and disposing of it.[6] However, because CTL fuels produced at plants without CCS have approximately twice the CO_2 emissions as petroleum-based fuels on a full-life-cycle basis, the unit cost of CTL when emitting the process CO_2 grows quickly as the CO_2-emission cost rises. With low CTL and CCS technology costs, applying CCS is more cost-effective for the producer than paying the CO_2-emission cost, even if that cost is as low as $10/ton CO_2. Moreover, in this situation, the low-cost CTL with CCS is cost-competitive with conventional fuels, given oil prices at or even somewhat below the 2007 EIA reference-case price.

With high costs for CTL and CCS, in comparison, CO_2-emission costs above about $30/ton will make investment in CCS more cost-effective than paying the CO_2-emission cost. With an emission cost this large, however, Figure 6.3 shows that the oil price would have to be above the EIA 2007 reference-case level for CTL with CCS to be cost-competitive with conventional fuels. CTL with CCS would be competitive with oil prices at or somewhat below the 2007 EIA high-price case.

Incorporating Energy-Security Costs

We noted at the start of this chapter that there is a considerable range of possible monetary benefits from expanding alternative-fuel supply in order to lower world oil prices. For EIA's reference-case oil price, we can look at a range of $4–$18/bbl. The upper end of this range reflects very optimistic assumptions (from the buyer's perspective) about the ability to overcome OPEC efforts to maintain above-competitive oil prices. The corresponding range with the high-cost-case EIA oil price is $6–$30/bbl. We can represent this energy-security premium on conventional oil in Figures 6.1–6.3 by considering upward shifts in the lines for conventional petroleum by the amounts just referenced. This incorporates an energy-security cost in the price of conventional fuel, just as we have incorporated the CO_2-emission costs in the full cost of the alternative fuels.

For oil sands, incorporating an energy-security premium into the cost of conventional petroleum does nothing to change the basic conclusion that SCO is already highly cost-competitive with conventional crude.[7] The implications for CTL in Figure 6.3 are somewhat more complicated. Relative to the EIA reference-case oil price in 2025, the range of the energy-security premium mentioned above is about $0.10–$0.45/gal. This premium improves

---

[6]   Consistent with the discussion in Chapters Three and Five, we are assuming that there are not retail-market sale opportunities for all the additional CO_2 generated.

[7]   One might want to argue that this premium should not apply to "friendly and reliable" conventional-oil imports from Canada. However, in a uniform global oil market, all oil suppliers benefit from the efforts by a subset of large producers to maintain above-competitive prices, and there is no way to shelter friendly suppliers from the impacts that alternative fuels would have on world oil prices.

the economics of CTL without CCS in the lower range of possible CO_2-emission costs. It also expands the range of possible CTL and CCS costs over which CTL with CCS would be economical. For the EIA high-oil-price case, the premium range is about \$0.15–0.70/gal. Adding this premium to the already higher cost of conventional fuel would only strengthen the cost-competitiveness of CTL.

CHAPTER SEVEN

Conclusions

In this final chapter, we first provide a synthesis of our analysis of the key influences on the cost-competitiveness of SCO and CTL relative to conventional petroleum. We then turn to broader conclusions and implications of the analysis.

Synthesis of the Cost-Competitiveness Analysis

Tables 7.1 and 7.2 show some of the key information from Figures 6.1 through 6.3 in Chapter Six. Table 7.1 identifies the highest cost of CO_2 emissions at which either SCO or CTL *without CCS* is cost-competitive with conventional petroleum, given different assumptions about oil prices (EIA's reference and high-price cases) and technology costs (base and high-price cases as presented in Chapters Four and Five). A key message from this comparison is that, although it is somewhat more CO_2-intensive than conventional petroleum, SCO has a cost advantage over conventional petroleum at a wide range of CO_2 prices. Only at prices that seem very high relative to what could be expected from actual CO_2-mitigation policies does the cost competition turn against SCO.

For CTL, in contrast, the degree of cost-competitiveness is more sensitive to the price of oil and the CO_2-emission cost. If CTL-production costs without CCS are well above our base-case assumptions, it will be cost-competitive only if oil prices are above the EIA reference-case price and if the cost of CO_2 emissions is not too high. This shows that, without CCS, CTL cost-competitiveness is vulnerable to CO_2 prices as well as to oil prices. With lower CTL costs, CTL is more competitive, though a CO_2-emission cost above \$30/ton and a return to more moderate longer-term oil prices would still cast a shadow over its competitiveness.

Table 7.2 focuses on how sensitive the alternatives *with* CCS are to changes in oil prices. With CCS, the relative cost of the alternatives and conventional petroleum are not very sensitive to the CO_2-emission cost, with one partial exception noted.

In Table 7.2, the second column shows the oil price below which the alternative is never competitive, even if production costs are low and there is no CO_2-emission cost. The entries for SCO indicate the robustness of its cost-competitiveness to crude-oil prices; only quite low long-term oil prices would lead to conventional oil being more cost-competitive. Again, since the CO_2 intensities of SCO and conventional oil do not differ much, these results are essentially independent of the assumed cost of CO_2 emissions.

The results for CTL in Table 7.2 show first that the threshold price above which oil must remain for CTL with CCS to be competitive is higher than for SCO. Moreover, if the costs of

Table 7.1
Influence of Carbon Dioxide–Emission Costs on the Competitiveness of Unconventional Fuels Compared to Conventional Petroleum, No Carbon Capture and Storage

Technology	Cost Scenario	Maximum CO_2-Emission Cost for Unconventional Fuel to Be Competitive ($/ton CO_2)	
		AEO 2007 Reference Case: $56/bbl crude, $1.74/gal diesel	AEO 2007 High-Oil-Price Case: $94/bbl crude, $2.51/gal diesel
SCO, integrated mining	Low	490	1,140
	High	240	630
SCO, in situ + upgrading	Low	310	770
	High	170	500
CTL	Base case	33	81
	High	0	47

Table 7.2
Sensitivity of Competitiveness of Unconventional Fuels with Carbon Capture and Storage to Crude-Oil Price

Technology	Price of Crude Oil Below Which Even Low-Cost Technology Is Never Economical ($/bbl)	Price of Crude Oil Above Which High-Cost Technology Can Be Economical ($/bbl)
SCO, integrated mining	31	40
SCO, in situ + upgrading	35	45
CTL	45	72

NOTE: To compare the production cost of FT diesel to petroleum, we assume that diesel sells for a 30-percent premium over that of low-sulfur, light crude oil, which is consistent with the assumption in Bartis, Camm, and Ortiz (forthcoming).

CTL with CCS are higher than our base case, then (as shown in Figure 6.3 in Chapter Six), the minimum oil price for CTL with CCS to be cost-competitive is higher as well.

Table 7.2 indicates that, with an oil price at or above $72/bbl, high-cost CTL plus CCS can be cost-competitive. At prices only slightly above $72/bbl, however, CTL with CCS will be competitive only with a very low CO_2-emission cost or success in getting CTL with CCS costs lower than our high-cost case.

Broader Conclusions and Implications

Even with future policy constraints on CO_2 emissions and their associated costs, SCO seems likely to be economically competitive with conventional petroleum unless future oil prices are relatively low. The main constraint on SCO appears to be its local and regional environmental impacts. We find that production of SCO from oil sands is likely to be cost-competitive with crude oil across a wide range of future oil prices and CO_2-emission costs. It is cost-competitive in particular if long-term future oil prices are at or above EIA's reference-case price projection and even for somewhat lower future oil prices. It is only about 15–20-percent more carbon-intensive on a life-cycle basis than conventional petroleum even without CCS.

Therefore, its competitiveness is not very sensitive to CO_2-emission costs.[1] Our calculations indicate relatively limited economic incentive to add CCS to SCO production unless future CO_2-emission costs are expected to be very high or CCS costs for oil-sand production are quite low.[2] The major factor potentially limiting the future growth of SCO production, at least in the absence of very stringent CO_2 limits, is likely to be the future availability of water as a key input and the environmental consequences of water use.

The economic competitiveness of coal-to-liquids is more dependent on future oil prices, carbon dioxide–sequestration costs, and the stringency of future carbon-dioxide limitations. CTL also has significant potential to be cost-competitive with conventional fuels over the longer term, but this depends on one of two distinct conditions being satisfied. The first condition is that either the future cost of CO_2 emissions or the cost of CCS is low. In that case, greater costs associated with mitigating or paying for CO_2 emissions do not offset CTL's potential cost advantage over conventional fuels. The second condition is that the price of crude oil in the longer term is significantly above EIA's 2007 reference level. This would make investment in CTL with CCS attractive.

Higher oil prices or significant energy-security premiums increase the economic desirability of synthetic crude oil and coal-to-liquids. If longer-term oil prices are high or future energy-security policy attaches a high premium to the market price of oil to account for energy-security costs, investment in both SCO and CTL production will be correspondingly more favorable. In particular, the magnitude of the CO_2-emission cost over which CTL without CCS is still economically attractive relative to conventional diesel will increase, and the economics of CTL with CCS can look attractive relative to conventional petroleum even if CCS turns out to be more costly. On the other hand, if oil prices end up being relatively low over the longer term, then CTL is less competitive than petroleum, even with a low CO_2-emission cost.

Unconventional fossil fuels do not, in themselves, offer a path to greatly reduced carbon-dioxide emissions, though there are additional possibilities for limiting emissions. Our analysis indicates that both CTL and SCO appear to offer economically competitive alternatives to conventional crude oil over a range of plausible future assumptions about oil prices and CCS costs. This means that they can be especially attractive opportunities for addressing the longer-term aspects of energy security—specifically, large wealth transfers resulting from uncompetitive oil pricing by foreign suppliers. Yet, even if successful on a large scale, applying CCS to producing CTL and SCO would still leave unaddressed the CO_2 emissions from final combustion of the fuels. Investments in expanding SCO or CTL do not, in themselves, offer a path toward the very large reductions in long-term CO_2 emissions from use of liquid fuels that would be needed to stabilize atmospheric concentrations of CO_2, a major consideration for those concerned with the long-term threats of climate change. Aside from some hypothetical future breakthrough in end-use capture and storage of CO_2 emissions, the path toward very low transportation-sector emissions is often seen to involve biofuels or very advanced electric-vehicle technologies. Another option, however, could be using CBTL *and* CCS in liquefaction plants (Bartis, Camm, and Ortiz, forthcoming). This possibility could make some CTL invest-

---

[1]   We noted also the sensitivity of SCO costs to natural-gas prices, but that sensitivity is not so large and high natural-gas prices would often be associated with high crude-oil prices.

[2]   One example of how this might be achieved is through a successful ICO_2N project in Alberta that provides low-cost access to CO_2 transportation and storage.

ment that is at least CCS ready an attractive option for expanding alternative fuels while maintaining options for significantly lower future CO_2 emissions—though several hurdles would need to be overcome in providing adequate and affordable biomass feedstock and in large-scale implementation of CCS. Initially, new CTL facilities with CCS may be able to avail themselves of a market for their CO_2 via increases in its use for EOR. However, it seems likely that, at a production level of several million barrels of CTL fuels per day, the EOR market would be saturated, and additional CO_2 emissions would entail bearing either the added costs of CO_2 storage or the costs of other CO_2-mitigation and -offset measures.

Relationships among the uncertainties surrounding oil prices, energy security, sequestration costs, and carbon dioxide–control stringency have important policy and investment implications for coal-to-liquids. Our analysis indicates that adding CCS to CTL is a good hedge against the cost of future CO_2 limitations if CCS can be realized on an adequately large scale, if CTL and CCS costs are in the lower part of the range of costs we have considered, and if future oil prices do not fall well below reference levels. If CTL and CCS costs are higher, however, the value to the CTL supplier of adding CCS to hedge against a high cost of future CO_2 controls is positive only with higher long-term oil prices. From a societal perspective, it is desirable to reduce the need for bearing higher long-term costs of more-aggressive and -costly CO_2-emission reductions. On the other hand, nearer-term concerns about energy security could lead to a situation in which there is a desire to keep nearer-term CO_2 limitations relatively modest while putting more emphasis on significant CTL investments even if they were made without incorporating CCS.

Neither CTL investors nor policymakers have many options for reducing long-term oil-price uncertainty. Moreover, there is a risk to the economic value of CTL investment just from the possibility of relatively low long-term prices. On the other hand, policymakers do have options for reducing the uncertainties surrounding CTL and CCS costs. There is a large social benefit from government financing for both continued R&D for CCS *and* initial CCS-test investments at a commercial operating scale to further assess the technical and economic characteristics of CCS. This analysis parallels the argument in Bartis, Camm, and Ortiz (forthcoming), who recommended active but limited public support for informative initial-scale CTL facilities. Conversely, it may be very beneficial socially to delay a significant ramp-up in CTL production until the uncertainties surrounding CCS technology and CTL-production costs can be reduced. These observations reflect the importance of the argument of the National Commission on Energy Policy (2004) for a broad portfolio of technology-development initiatives, as well as a variety of policy instruments, to promote energy diversity and decarbonization of fuel sources.

References

Air Resources Board, "Low Carbon Fuel Standard Program," last reviewed June 5, 2008. As of June 27, 2008:
http://www.arb.ca.gov/fuels/lcfs/lcfs.htm

Alberta Chamber of Resources, *Oil Sands Technology Roadmap: Unlocking the Potential*, Edmonton, Alta., 2004. As of August 5, 2008:
http://www.acr-alberta.com/OSTR_report.pdf

Alberta Employment, Immigration and Industry, *Alberta Oil Sands Industry Update*, June 2007. As of August 6, 2008:
http://www.albertacanada.com/documents/AIS-EC_oilSandsUpdate_1207.pdf

Alberta Energy Resources Conservation Board (formerly the Alberta Energy and Utilities Board), *Alberta's Energy Reserves 2007 and Supply/Demand Outlook 2008–2017*, Calgary, Alta., 2008.

Alberta Environment, "Athabasca River Water Management Framework," undated Web page. As of June 17, 2008:
http://environment.alberta.ca/1546.html

———, "Water in the Oil Sands Industry," briefing to CONRAD Oil Sands Water Usage Workshop, Fort McMurray, Alta., February 2004. As of June 17, 2008:
http://www.conrad.ab.ca/seminars/water_usage/2004/Water_in_the_oil_sands_industry_Marriott.pdf

———, "Alberta Issues First-Ever Oil Sands Land Reclamation Certificate," press release, Edmonton, Alta., March 19, 2008. As of June 17, 2008:
http://www.alberta.ca/home/NewsFrame.cfm?ReleaseID=/acn/200803/23196C8880E90-A0E1-9CE0-1B3799BC38A51E3E.html

Albian Sands, "Athabasca Oil Sands Project," undated Web page. As of June 17, 2008 (follow Athabasca Oil Sands Project link):
http://www.albiansands.ca/environment.htm

Amos, W., *Summary of Chariton Valley Switchgrass Co-Fire Testing at the Ottumwa Generating Station in Chillicothe, Iowa: Milestone Completion Report*, Golden, Colo.: National Renewable Energy Laboratory, NREL/TP-510-32424, 2002.

Argonne National Laboratory, "The Greenhouse Gases, Regulated Emissions, and Energy Use in Transportation (GREET) Model," version 1.8a, 1999. As of August 29, 2008:
http://www.transportation.anl.gov/modeling_simulation/GREET/

Argote, Linda, and Dennis Epple, "Learning Curves in Manufacturing," *Science*, Vol. 247, No. 4945, February 23, 1990, pp. 920–924.

Baker, John M., Tyson E. Ochsner, Rodney T. Venterea, and Timothy J. Griffis, "Tillage and Soil Carbon Sequestration: What Do We Really Know?" *Agriculture Ecosystems and Environment*, Vol. 118, 2007, pp. 1–5.

Bartis, James T., *Policy Issues for Coal-to-Liquid Development*, testimony before the U.S. Senate Committee on Energy and Natural Resources, Santa Monica, Calif.: RAND Corporation, CT-281, May 24, 2007. As of June 17, 2008:
http://www.rand.org/pubs/testimonies/CT281/

Bartis, James T., Frank Camm, and David S. Ortiz, *Producing Liquid Fuels from Coal: Prospects and Policy Issues*, Santa Monica, Calif.: RAND Corporation, MG-754-AF/NETL, forthcoming.

Bartis, James T., Tom LaTourrette, Lloyd Dixon, D. J. Peterson, and Gary Cecchine, *Oil Shale Development in the United States: Prospects and Policy Issues*, Santa Monica, Calif.: RAND Corporation, MG-414-NETL, 2005. As of June 16, 2008:
http://www.rand.org/pubs/monographs/MG414/

BEA—*see* Bureau of Economic Analysis.

Bergerson, Joule, "Inquiry Regarding Oil Sands Data and Information," phone conversation with Aimee Curtright, October 23, 2007.

Bergerson, Joule, and David Keith, *Life Cycle Assessment of Oil Sands Technologies*, Calgary, Alta.: Institute for Sustainable Energy, Environment and Economy, Alberta Energy Futures Project paper 11, November 2006. As of June 17, 2008:
http://www.iseee.ca/files/iseee/ABEnergyFutures-11.pdf

BLM—*see* U.S. Department of Interior, Bureau of Land Management.

BLS—*see* Bureau of Labor Statistics.

Bordetsky, Ann, *Driving It Home: Choosing the Right Path for Fueling North America's Transportation Future*, New York: National Resources Defense Council, June 2007. As of June 17, 2008:
http://www.nrdc.org/energy/drivingithome/contents.asp

Buchanan, T., R. Schoff, and J. White, *Updated Cost and Performance Estimates for Fossil Fuel Power Plants with CO_2 Removal*, Palo Alto, Calif.: Electronic Power Research Institute, interim report 1004483, December 2002. As of June 17, 2008:
http://www.netl.doe.gov/technologies/carbon_seq/Resources/Analysis/pubs/UpdatedCosts.pdf

Bureau of Economic Analysis, "Table 1.1.4. Price Indexes for Gross Domestic Product, Seasonally Adjusted," U.S. Department of Commerce, 2008. As of August 5, 2008:
http://www.bea.gov/bea/dn/nipaweb/TableView.asp?SelectedTable=4&FirstYear=2005&LastYear=2007&Freq=Qtr

Bureau of Labor Statistics, "Producer Price Index Industry Data," series pcu324110324110, industry: petroleum refineries, product: petroleum refineries, base date: 198506, undated data retrieved.

California Air Resources Board, "Low Carbon Fuel Standard Program," Web page, last reviewed July 29, 2008. As of July 29, 2008:
http://www.arb.ca.gov/fuels/lcfs/lcfs.htm

Canadian Oil Sands Trust, "At a Glance," undated Web page. As of August 25, 2008:
http://www.cos-trust.com/asset/

CARB—*see* California Air Resources Board.

Crutzen, P. J., A. R. Mosier, K. A. Smith, and W. Winiwarter, "N_2O Release from Agro-Biofuel Production Negates Global Warming Reduction by Replacing Fossil Fuels," *Atmospheric Chemistry and Physics Discussions*, Vol. 7, August 2007, pp. 11191–11205.

DKRW Advanced Fuels, "Medicine Bow Fuel & Power LLC," undated Web page. As of February 21, 2008:
http://www.dkrwadvancedfuels.com/fw/main/Medicine_Bow-111.html

DOE—*see* U.S. Department of Energy.

EIA—*see* Energy Information Administration.

Energy Information Administration, *Annual Coal Report 2005*, Washington, D.C., DOE/EIA-0584(2005), October 2006. As of June 17, 2008:
http://tonto.eia.doe.gov/FTPROOT/coal/05842005.pdf

———, *Annual Energy Outlook 2007 with Projections to 2030*, Washington, D.C., DOE/EIA-0383(2007), February 2007a. As of June 16, 2008:
http://www.eia.doe.gov/oiaf/archive/aeo07/

———, *International Energy Outlook 2007*, Washington, D.C., DOE/EIA-0484(2007), May 2007b. As of August 5, 2008:
http://www.eia.doe.gov/oiaf/archive/ieo07/

———, *Annual Energy Review 2006*, Washington, D.C., DOE/EIA-0384(2006), June 26, 2007c. As of June 17, 2008:
http://tonto.eia.doe.gov/FTPROOT/multifuel/038406.pdf

———, "U.S. Primary Energy Consumption by Source and Sector, 2006 (Quadrillion Btu)," June 27, 2007d.

———, *Emissions of Greenhouse Gases Report*, Washington, D.C., DOE/EIA-0573(2006), November 28, 2007e. As of June 17, 2008:
http://www.eia.doe.gov/oiaf/1605/ggrpt/

———, *Energy Market and Economic Impacts of S.2191, the Lieberman-Warner Climate Security Act of 2007*, Washington, D.C.: U.S. Department of Energy, 2008a.

———, *International Energy Outlook 2008: Highlights*, Washington, D.C., DOE/EIA-0484(2008), June 2008b. As of August 10, 2008:
http://www.eia.doe.gov/oiaf/ieo/

———, "U.S. Natural Gas Wellhead Price," Washington, D.C., last updated July 29, 2008c. As of August 5, 2008:
http://tonto.eia.doe.gov/dnav/ng/hist/n9190us3m.htm

———, "Cushing, OK WTI Spot Price FOB," Washington, D.C., last updated August 6, 2008d. As of August 5, 2008:
http://tonto.eia.doe.gov/dnav/pet/hist/rwtcM.htm

Energy Resources Conservation Board, *Alberta's Energy Reserves 2006 and Supply/Demand Outlook, 2007–2016*, Calgary, Alta.: Alberta Energy and Utilities Board, ST98-2007, June 2007. As of June 17, 2008:
http://www.ercb.ca/docs/products/STs/st98-2007.pdf

Environment Canada, "Information on Greenhouse Gas Sources and Sinks, 2006 Emissions Data, Table 3: Summary of GHG Emissions by Facility," updated August 1, 2007a. As of August 6, 2008:
http://www.ec.gc.ca/pdb/ghg/onlinedata/DataAndReports_e.cfm

———, "Welcome to the Western Boreal Conservation Initiative," last updated October 3, 2007b. As of June 17, 2008:
http://www.mb.ec.gc.ca/nature/ecosystems/wbci-icbo/index.en.html

———, "2006 Emissions Data: Table 3: Summary of GHG Emissions by Facility," last updated June 17, 2008. As of June 17, 2008:
http://www.ec.gc.ca/pdb/ghg/onlinedata/kdt_t3_e.cfm?year=2006

ERCB—see Energy Resources Conservation Board.

Fargione, Joseph, Jason Hill, David Tillman, Stephen Polasky, and Peter Hawthorne, "Land Clearing and the Biofuel Carbon Debt," *Science Express*, Vol. 319, February 7, 2008, pp. 1235–1238.

Farrell, Alex, and Michael O'Hare, "Greenhouse Gas (GHG) Emissions from Indirect Land Use Change (LUC)," memorandum to California Air Resources Board, Berkeley, Calif.: Energy and Resources Group, University of California, Berkeley, 2008.

Government of Alberta, "Out of the Air and into the Ground: Alberta and Canada Join Forces to Assess Technology to Capture Greenhouse Gases," press release, March 8, 2007. As of July 29, 2008:
http://www.gov.ab.ca/acn/200703/2114533C2D9BD-0A8C-A2FE-37510D4AE5562B1D.html

———, "Alberta Issues First-Ever Oil Sands Land Reclamation Certificate: Former Oil Sands Site Transformed into Forested Area," press release, March 19, 2008a. As of July 29, 2008:
http://www.alberta.ca/home/NewsFrame.cfm?ReleaseID=/acn/200803/23196C8880E90-A0E1-9CE0-1B3799BC38A51E3E.html

————, "Legislation Launches Climate Change Fund as Vehicle to Deliver Real Emission Reductions: Amendment Ensures Strategic Investment and Accountability," press release, April 30, 2008b. As of July 29, 2008:
http://alberta.ca/home/NewsFrame.cfm?ReleaseID=/acn/200804/23419A1030535-DBAD-651B-8FB5E6FBDB74469C.html

————, "Alberta Surges Ahead with Climate Change Action Plan: $2 Billion Invested in Carbon Capture and Storage; $2 billion in Public Transit," press release, July 8, 2008c. As of July 29, 2008:
http://alberta.ca/home/NewsFrame.cfm?ReleaseID=/acn/200807/23960039FB54D-CC21-7234-31C3E853089A1E6C.html

Gray, David, and Charles White, telephone communication with David Ortiz, July 23, 2007.

Griffiths, Mary, Dan Woynillowicz, and Amy Taylor, *Troubled Waters, Troubling Trends: Technology and Policy Options to Reduce Water Use in Oil and Oil Sands Development in Alberta*, Drayton Valley, Alta.: Pembina Institute, 2006. As of June 17, 2008:
http://www.pembina.org/pub/612

Hess, Ronald Wayne, and Christopher W. Myers, *Assessing Initial-Cost Growth and Subsequent Long-Term Cost Improvement with Application to Coal-to-SNG Processes*, Santa Monica, Calif.: RAND Corporation, N-2783-GRI, August 1988.

Huntington, Hillard G., *The Economic Consequences of Higher Crude Oil Prices*, Stanford, Calif.: U.S. Department of Energy, Energy Modeling Forum final report SR 9, October 2005. As of June 17, 2008:
http://www.stanford.edu/group/EMF/publications/doc/EMFSR9.pdf

ICO_2N, undated home page. As of July 29, 2008:
http://www.ico2n.com/

IPCC—*see* Intergovernmental Panel on Climate Change.

Intergovernmental Panel on Climate Change, *IPCC Special Report on Carbon Dioxide Capture and Storage*, Bert Metz, Ogunlade Davidson, Heleen de Coninck, Manuela Loos, and Leo Meyer, eds., New York: Cambridge University Press, 2005.

————, *Climate Change 2007: Synthesis Report. Contribution of Working Groups I, II and III to the Fourth Assessment Report of the Intergovernmental Panel on Climate Change*, R. K. Pachauri and A. Reisinger eds., Geneva, Switzerland, 2007.

Kinder Morgan, *Kinder Morgan Inc. 2005 Annual Report*, Houston, Tex.: Kinder Morgan, Inc., 2006. As of October 5, 2007:
http://www.kindermorgan.com/investor/kmi_2005_annual_report_overview.pdf

Kuuskraa, Vello A., "CO_2-EOR: An Enabling Bridge for the Oil Transition," *Modeling the Oil Transition: A Summary of the Proceedings of the DOE/EPA Workshop on the Economic and Environmental Implications of Global Energy Transition*, Washington, D.C., April 20–21, 2006.

Lacombe, Romain H., and John E. Parsons, *Technologies, Markets and Challenges for Development of the Canadian Oil Sands Industry*, Cambridge, Mass.: Center for Energy and Environmental Policy Research, working paper 07-006, June 2007. As of June 17, 2008:
http://web.mit.edu/ceepr/www/publications/workingpapers/2007-006.pdf

Landsberg, Hans H., and Kenneth Joseph Arrow, *Energy: The Next Twenty Years*, Cambridge, Mass.: Ballinger Pub. Co., 1979.

Leiby, Paul N., *Estimating the Energy Security Benefits of Reduced U.S. Oil Imports*, Oak Ridge, Tenn.: Oak Ridge National Laboratory, ORNL/TM-2007/028, July 2007. As of June 17, 2008:
http://www.epa.gov/otaq/renewablefuels/ornl-tm-2007-028.pdf

Lieberman, M. B., "The Learning-Curve and Pricing in the Chemical-Processing Industries," *RAND Journal of Economics*, Vol. 15, No. 2, 1984, pp. 213–228.

Massachusetts Institute of Technology, *The Future of Coal: Options for a Carbon-Constrained World*, Cambridge, Mass., 2007. As of June 23, 2008:
http://web.mit.edu/coal/

McCulloch, Matthew, Marlo Raynolds, and Rich Wong, *Carbon Neutral 2020: A Leadership Opportunity in Canada's Oil Sands*, Drayton Valley, Alta.: Pembina Institute, oil sands issue paper 2, October 2006. As of June 17, 2008:
http://www.pembina.org/pub/1316

Merrow, Edward W., *An Analysis of Cost Improvement in Chemical Process Technologies*, Santa Monica, Calif.: RAND Corporation, R-3357-DOE, 1989. As of August 6, 2008:
http://www.rand.org/pubs/reports/R3357/

Merrow, Edward W., Kenneth Phillips, and Christopher W. Myers, *Understanding Cost Growth and Performance Shortfalls in Pioneer Process Plants*, Santa Monica, Calif.: RAND Corporation, R-2569-DOE, 1981. As of June 17, 2008:
http://www.rand.org/pubs/reports/R2569/

Milbrandt, Anelia, *A Geographic Perspective on the Current Biomass Resource Availability in the United States*, Golden, Colo.: National Renewable Energy Laboratory, December 2005. As of June 17, 2008:
http://www.nrel.gov/docs/fy06osti/39181.pdf

MIT—*see* Massachusetts Institute of Technology.

Myers, Christopher W., and R. Ylmaz Arguden, *Capturing Pioneer Plant Experience: Implications for Synfuel Projects*, Santa Monica, Calif.: RAND Corporation, N-2063-SFC, 1984. As of June 17, 2008:
http://www.rand.org/pubs/notes/N2063/

Myers, Christopher W., Ralph F. Shangraw, Mark R. Devey, and Toshi Hayashi, *Understanding Process Plant Schedule Slippage and Startup Costs*, Santa Monica, Calif.: RAND Corporation, R-3215-PSSP/RC, 1986. As of June 17, 2008:
http://www.rand.org/pubs/reports/R3215/

Nakamura, David, "Special Report: Canadian, US Processors Adding Capacity to Handle Additional Oil Sands Production," *Oil and Gas Journal*, Vol. 105, No. 26, July 9, 2007, pp. 54–58.

National Coal Council, *Coal: America's Energy Future*, Vol. I, Washington, D.C., 2006. As of August 5, 2008:
http://www.coalamericasenergyfuture.com/pdf/NCC_Report_Vol1.pdf

National Commission on Energy Policy, *Ending the Energy Stalemate: A Bipartisan Strategy to Meet America's Energy Challenges*, Washington, D.C., December 2004. As of August 28, 2008:
http://www.energycommission.org/files/finalReport/O82F4692.pdf

National Energy Board, *Canada's Oil Sands: Opportunities and Challenges to 2015*, Calgary, Alta., 2004.

———, *Canada's Oil Sands: Opportunities and Challenges to 2015: An Update*, Calgary, Alta., 2006.

———, "Inquiry Regarding Update to 'Canada's Oil Sands' Report," email correspondence with board staff, July 11, 2008.

National Energy Technology Laboratory, "Gasification Database," undated Web page. As of August 5, 2008:
http://www.netl.doe.gov/technologies/coalpower/gasification/database/database.html

———, *Beluga Coal Gasification Feasibility Study: Phase I Final Report*, DOE/NETL-2006/1248, July 2006. As of June 17, 2008:
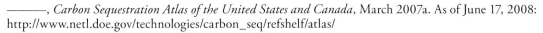
http://204.154.137.14/technologies/coalpower/gasification/pubs/pdf/Beluga%20Coal%20Gasif%20Feasibility%20Study9_15_06.pdf

———, *Carbon Sequestration Atlas of the United States and Canada*, March 2007a. As of June 17, 2008:
http://www.netl.doe.gov/technologies/carbon_seq/refshelf/atlas/

———, *Carbon Sequestration Technology Roadmap and Program Plan 2007*, April 2007b. As of June 17, 2008:
http://www.netl.doe.gov/technologies/carbon_seq/refshelf/project%20portfolio/2007/2007Roadmap.pdf

———, *Baseline Technical and Economic Assessment of a Commercial Scale Fischer-Tropsch Liquids Facility*, Pittsburgh, Pa., DOE/NETL-2007/1260, April 9, 2007c. As of June 17, 2008:
http://204.154.137.14/energy-analyses/pubs/Baseline%20Technical%20and%20Economic%20Assessment%20of%20a%20Commercial%20S.pdf

———, *Increasing Security and Reducing Carbon Emissions of the U.S. Transportation Sector: A Transformational Role for Coal with Biomass*, DOE/NETL-2007/1298, August 24, 2007d. As of June 17, 2008: http://204.154.137.14/energy-analyses/pubs/NETL-AF%20CBTL%20Study%20Final%202007%20 Aug%2024.pdf

———, *Storing CO$_2$ with Enhanced Oil Recovery*, Pittsburgh, Pa., DOE/NETL-402/1312/02-07-08, February 7, 2008. As of June 17, 2008: http://www.netl.doe.gov/energy-analyses/pubs/storing%20co2%20w%20eor_final.pdf

National Petroleum Council, *Hard Truths: Facing the Hard Truths About Energy: A Comprehensive View to 2030 of Global Oil and Natural Gas*, Washington, D.C., July 2007. As of June 17, 2008: http://downloadcenter.connectlive.com/events/npc071807/pdf-downloads/NPC_Facing_Hard_Truths.pdf

National Research Council, *Coal: Research and Development to Support National Energy Policy*, Washington, D.C.: National Academies Press, 2007. As of June 17, 2008: http://www.nap.edu/books/030911022X/html/

NCEP—*see* National Commission on Energy Policy.

NEB—see National Energy Board.

NETL—*see* National Energy Technology Laboratory.

NPC—*see* National Petroleum Council.

Ohio River Clean Fuels, press release, Vancouver, Wash.: Baard Energy, Inc., September 26, 2007.

Oil Shale and Tar Sands Programmatic Environmental Impact Statement Information Center, "About Tar Sands," undated Web page. As of June 17, 2008: http://ostseis.anl.gov/guide/tarsands/

OPTI Canada, *Delivering Our Oil Sands Advantage: Annual Report 2006*, Calgary, Alta., 2007.

Parry, Ian W. H., and Joel Darmstadter, *The Costs of U.S. Oil Dependency*, Washington, D.C.: Resources for the Future, discussion paper 03-59, December 2003. As of June 17, 2008: http://www.rff.org/Documents/RFF-DP-03-59.pdf

Perlack, Robert D., Lynn L. Wright, Anthony F. Turhollow, Robin L. Graham, Bryce J. Stokes, and Donald C. Erbach, *Biomass as Feedstock for a Bioenergy and Bioproducts Industry: The Technical Feasibility of a Billion-Ton Annual Supply*, Oak Ridge, Tenn.: Oak Ridge National Laboratory, 2005. As of June 17, 2008: http://feedstockreview.ornl.gov/pdf/billion%5Fton%5Fvision.pdf

Platts Global Power Report, "Bruce Power Signs Deal to Acquire Assets of Energy Alberta, Including Nuclear Project," December 6, 2007, North America: Secondary Markets section, p. 17.

Public Law 96-294, Energy Security Act, June 30, 1980.

Public Law 109-58, Energy Policy Act of 2005, August 8, 2005.

Public Law 110-140, Energy Independence and Security Act, December 19, 2007.

Rentech, Inc., "Rentech Projects," 2008. As of February 21, 2008: http://www.rentechinc.com/rentech-projects.htm#2

Sasol Limited, *Reaching New Energy Frontiers Through Competitive GTL Technology*, corporate brochure, Johannesburg, June 2006.

———, *Project Update: Oryx Gas-to-Liquids (GTL) Joint Venture*, Johannesburg, 2007a.

———, *Investor Insight*, Johannesburg, May 22, 2007b.

Schurr, Sam H., Joel Darmstadter, Harry Perry, William Ramsey, and Milton Russell, *Energy in America's Future: The Choices Before Us*, Baltimore, Md.: Johns Hopkins University Press for Resources for the Future, 1979.

Searchinger, Timothy, Ralph Heimlich, R. A. Houghton, Fengxia Dong, Amani Elobeid, Jacinto Fabiosa, and Simla Tokgoz, "Use of U.S. Croplands for Biofuels Increases Greenhouse Gases Through Emissions from Land Use Change," *Science Express*, Vol. 319, February 7, 2008, pp. 1238–1240.

Southern States Energy Board, *American Energy Security: Building a Bridge to Energy Independence and to a Sustainable Energy Future*, Norcross, Ga., July 2005. As of June 23, 2008:
http://www.americanenergysecurity.org/studyrelease.html

Speight, J. G., *The Chemistry and Technology of Petroleum*, 4th ed., Boca Raton, Fla.: CRC Press/Taylor and Francis, 2007.

SSEB—*see* Southern States Energy Board.

Steynberg, Andre, *Coal-to-Liquids: An Alternative Oil Supply?* workshop report of the November 2, 2006, meeting of the Coal Industry Advisory Board, Paris, 2006.

Strategy West, *Existing and Proposed Canadian Commercial Oil Sands Projects*, April 2008. As of June 17, 2008:
http://www.strategywest.com/downloads/StratWest_OSProjects.pdf

Suncor, "About Suncor," undated Web page. As of June 17, 2008:
http://www.suncor.com/default.aspx?ID=1

———, *2007 Report on Sustainability: A Closer Look at Our Journey Toward Sustainable Development*, Calgary, Alta., 2007a. As of July 29, 2008:
http://www.suncor.com/data/1/rec_docs/1398_SD%20Report%202007(new)white.pdf

———, "Suncor Energy Reports Oil Sands Production Numbers for December 2006," press release, Calgary, Alta., January 4, 2007b. As of June 17, 2008:
http://www.suncor.com/default.aspx?ID=2897

———, *Climate Change: A Decade of Taking Action: 2007 Progress Report on Climate Change*, Calgary, Alta., September 2007c. As of June 17, 2008:
http://www.suncor.com/data/1/rec_docs/1481_Suncor%202007%20Climate%20Change%20Report.pdf

Syncrude, "Oil Sands History," undated Web page. As of August 25, 2008:
http://www.syncrude.ca/users/folder.asp?FolderID=5657

———, *2006 Sustainability Report*, Fort McMurray, Alta., c. 2006. As of June 17, 2008:
http://sustainability.syncrude.ca/sustainability2006/download/SyncrudeSD2006.pdf

Task Force on Strategic Unconventional Fuels, *America's Strategic Unconventional Fuels: Oil Shale, Tar Sands, Coal Derived Liquids, Heavy Oil, CO_2 Enhanced Recovery and Storage*, Vol. III: *Resource and Technology Profiles*, Washington, D.C., 2007.

Tilman, David, Jason Hill, and Clarence Lehman, "Carbon-Negative Biofuels from Low-Input High-Diversity Grassland Biomass," *Science*, Vol. 314, December 8, 2006, pp. 1598–1600.

Timilsina, Govinda R., Nicole LeBlanc, and Thorn Walden, *Economic Impacts of Alberta's Oil Sands*, Calgary, Alta.: Canadian Energy Research Institute, Vol. 1, study 110, October 2005. As of June 17, 2008:
http://www.ceri.ca/Publications/documents/OilSandsReport-Final.PDF

Toman, Michael, James Griffin, and Robert J. Lempert, *Impacts on U.S. Energy Expenditures and Greenhouse-Gas Emissions of Increasing Renewable-Energy Use*, Santa Monica, Calif.: RAND Corporation, TR-384-1-EFC, 2008. As of August 8, 2008:
http://www.rand.org/pubs/technical_reports/TR384-1/

TransAlta, "TransAlta Signs Agreement with Technology Partner Alstom to Develop Carbon Capture and Storage Project in Alberta, Canada," press release, April 3, 2008. As of July 29, 2008:
http://www.transalta.com/transalta/webcms.nsf/AllDoc/2D8D876D809831848725741F0076700F?OpenDocument

U of U—*see* University of Utah, Institute for Clean and Secure Energy, Utah Heavy Oil Program.

U.S. Department of Energy, "Bioenergy," undated. As of October 5, 2007:
http://www.doe.gov/energysources/bioenergy.htm

———, Office of Petroleum Reserves, Office of Naval Petroleum and Oil Shale Reserves, *Secure Fuels from Domestic Resources: The Continuing Evolution of America's Oil Shale and Tar Sands Industries*, Washington, D.C., June 2007a. As of July 29, 2008:
http://www.fossil.energy.gov/programs/reserves/npr/Secure_Fuels_from_Domestic_Resources_-_P.pdf

U.S. Department of Interior, Bureau of Land Management, *Oil Shale and Tar Sands Draft Programmatic Environmental Impact Statement*, Washington, D.C., DES 07-60, December 2007. As of August 5, 2008:
http://ostseis.anl.gov/eis/guide/

U.S. House of Representatives, Findings and Recommendations of the Advisory Panel on Synthetic Fuels, Report to the Committee on Science and Technology, 96th Congress, 2nd Session, August 1980.

U.S. Statutes, Title 58, Section 189, Synthetic Liquid Fuels Act, April 5, 1944.

U.S. Geological Survey, *Natural Bitumen Resources of the United States*, Reston, Va., December 2006. As of June 17, 2008:
http://pubs.usgs.gov/fs/2006/3133/

USGS—*see* U.S. Geological Survey.

University of Utah, Institute for Clean and Secure Energy, Utah Heavy Oil Program, *A Technical, Economic, and Legal Assessment of North American Heavy Oil, Oil Sands, and Oil Shale Resources*, September 2007. As of August 5, 2008:
http://www.fossil.energy.gov/programs/oilgas/publications/oilshale/HeavyOilLowRes.pdf

van Dongen, Ad, and Marco Kanaar, *Co-Gasification at the Buggenum IGCC Power Plant*, Buggenum, Netherlands: NUON Power, 2006.

Williams, Robert H., Eric D. Larson, and Haiming Jin, *Synthetic Fuels in a World with High Oil and Carbon Prices*, prepared for Eighth International Conference on Greenhouse Gas Control Technologies, Trondheim, Norway, June 19–22, 2006. As of June 17, 2008:
http://www.futurecoalfuels.org/documents/032007_williams.pdf

Woynillowicz, Dan, Chris Severson-Baker, and Marlo Raynolds, *Oil Sands Fever: The Environmental Implications of Canada's Oil Sands Rush*, Pembina Institute, November 2005. As of June 17, 2008:
http://pubs.pembina.org/reports/OSF_Fact72.pdf

Zwart, Robin W. R., and Harold Boerrigter, "High Efficiency Co-Production of Synthetic Natural Gas (SNG) and Fischer-Tropsch (FT) Transportation Fuels from Biomass," *Energy and Fuels*, Vol. 19, No. 2, March 2005, pp. 591–597.